The Cambridge Manuals of Science and
Literature

NATURAL SOURCES OF ENERGY

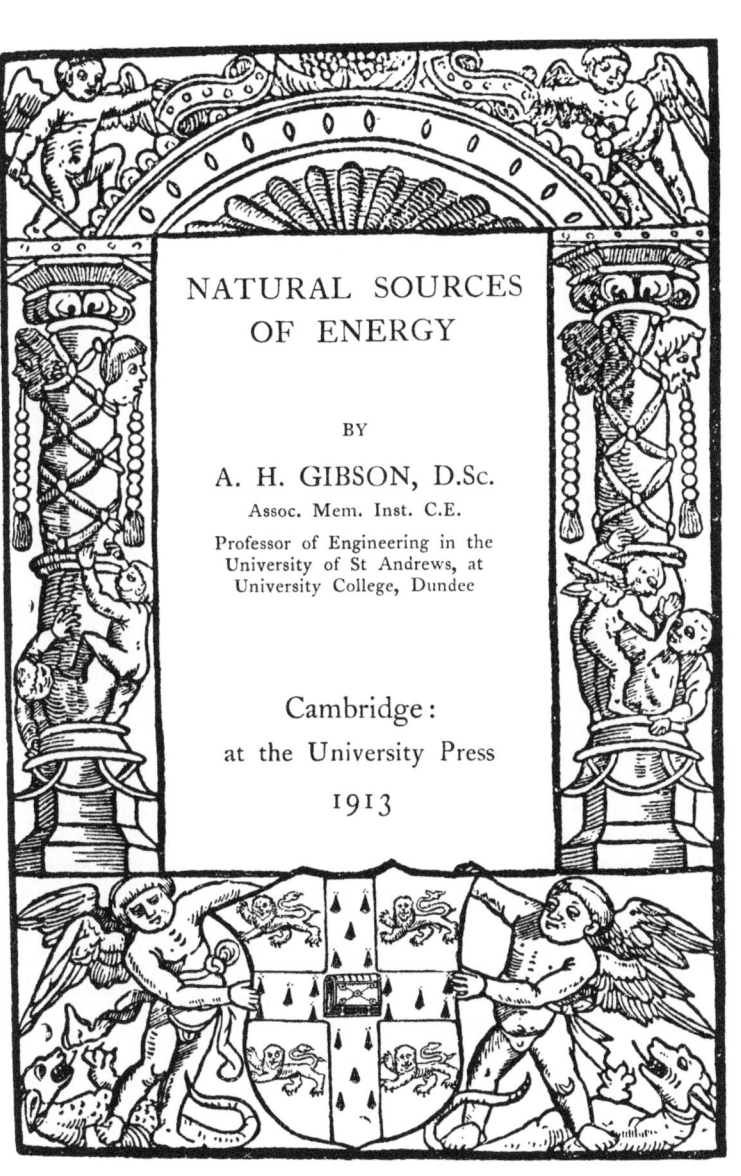

NATURAL SOURCES OF ENERGY

BY

A. H. GIBSON, D.Sc.

Assoc. Mem. Inst. C.E.

Professor of Engineering in the
University of St Andrews, at
University College, Dundee

Cambridge:

at the University Press

1913

CAMBRIDGE UNIVERSITY PRESS
Cambridge, New York, Melbourne, Madrid, Cape Town,
Singapore, São Paulo, Delhi, Tokyo, Mexico City

Cambridge University Press
The Edinburgh Building, Cambridge CB2 8RU, UK

Published in the United States of America by
Cambridge University Press, New York

www.cambridge.org
Information on this title: www.cambridge.org/9781107401730

© Cambridge University Press 1913

First published 1913
First paperback edition 2011

A catalogue record for this publication is available from the British Library

ISBN 978-1-107-40173-0 Paperback

*With the exception of the coat of arms
at the foot, the design on the title page is a
reproduction of one used by the earliest known
Cambridge printer, John Siberch, 1521*

PREFACE

MUCH attention has been paid, of recent years, to the rapid depletion of the coal resources of our own and other actively industrial countries, and to the problems which this depletion must ultimately involve. Estimates of the probable length of time for which such resources will still remain available vary considerably. The pessimist says a few—very few—hundred years; the optimist many thousands of years; and opinions as to the feasibility of replacing such stores of energy, when exhausted, by energy derived from other sources are as widely divergent.

From the very nature of the case it is impossible to do more than give an intelligent guess at the magnitude of the remaining deposits of fossil fuels. Still recent investigations in this and in other countries have thrown a good deal of light on the subject, and it is probable that the latest estimates are reasonably close to the truth. Also our knowledge of other natural sources of energy is now sufficiently complete to enable us to obtain some fair idea of their possibilities as factors in any general power scheme.

In view of the remarkable discoveries made in
the domain of molecular physics during the past
decade, it is indeed conceivable that the ultimate
solution of the world's energy problem may be independent of those natural agencies with which the
present age is familiar, and that the engineer of the
future may have at his disposal, and be able to
utilize and control molecular activities such as are,
at present, entirely beyond our reach. Such a solution is, however, even beyond the scope of rational
speculation in our present state of knowledge, and
while not impossible, it would appear to be in a high
degree improbable.

Nor, fortunately, is there any reason to anticipate
that the utilization of such sources of energy will
be necessary in order to maintain, indefinitely, the
normal industrial activities of the world. Those
agencies with which we are at present more or less
familiar are full of possibilities, and the following
pages are devoted to a consideration of these possibilities and of their bearing on the probable solution
of the energy problem.

A. H. G.

DUNDEE.
   July, 1913.

CONTENTS

LIST OF ILLUSTRATIONS

CHAPTER I

To most minds there is a certain fascination in the attempt to forecast something of the conditions of life and social activity on earth through the centuries ahead. For us, living in an age throughout which the growth of scientific knowledge and mechanical invention has been continuous and has achieved results beyond the dreams of our predecessors, there is a natural tendency to assume that such progress will be continuous in the future as in the immediate past, and that the coming centuries will as far exceed our own in mechanical, industrial, and inventive activity as this exceeds the centuries gone by. Whether such an assumption is justifiable or not, only time can tell. Periods of relatively intense activity and material progress have occurred from time to time in the past, notably in the civilizations of ancient Egypt, of Babylonia, Greece and Rome, but only to be succeeded by periods of stagnation in which much of the previous gain has been lost. In this, as in

other respects, it is not inconceivable that History
may repeat itself. Still, there is a fundamental difference between the world-wide distribution of our
present civilization and the local nature of those
which have preceded it, and it is all but unthinkable
that any such relapse as has occurred in the past can
again be possible.

Whether the happiness and well-being of mankind
would not indeed be better conserved by a general
slackening of the intense activity and competition of
the present age is another matter ; but whatever the
future may hold, there is little sign of any tendency
in this direction at present, and we must face the fact
that a considerable increase in material progress and
industrial activity in the years to come is not only
possible but extremely probable.

And, granting this, we are bound to recognise
that it must involve immense changes in many of
the conditions which we are apt to regard as a settled
feature of the present or of any future dispensation.
Social and economic forces tending to this end are
continuously at work, and, among the latter, one of
the most important is that due to the gradual depletion of the world's store of cheaply available
energy.

It is axiomatic that the permanence, still more
the growth, of the present state of material activity
depends entirely upon the availability of an ample

supply of mechanical energy. Moreover, everything points to the conclusion that it cannot be long— geologically speaking—before the most easily available of such supplies, those of our coal- and oil-fields, are so seriously depleted as to become a negligible factor in the general energy scheme of the globe.

As to their ultimate exhaustion there can be no doubt. Representing as they do the slowly accumulated savings of radiant energy poured on to the earth during countless thousands of years they are yet strictly limited in amount, and though that amount is in the aggregate enormous, the experience of recent years teaches that in our own portion of the globe they are even now beginning to show signs of the end.

Our knowledge of the magnitude of these deposits and of their probable duration is of necessity very incomplete. The first systematic investigation of the coal measures of this country was made by a Royal Commission appointed in 1866. After considering all the available evidence, they reported in 1871 to the effect that the amount of coal available in proved fields at depths less than 4000 feet was approximately 90,000 million tons, and that a further 56,000 million tons was available in fields not as yet tapped. A second Commission appointed in 1901 reported in 1905 to the effect that the coal available in proved fields was roughly 101,000 million tons ; in unproved fields,

39,400 million tons, with an additional 5200 million tons at depths greater than 4000 feet; giving a grand total of 145,600 million tons.

The total output of British collieries in 1870 was about 110 million tons. In 1905 it had risen to 236 million tons, and in 1911 to about 272 million tons. Thus, on the above estimates, if the present rate of output were to be maintained, and if the whole of the coal could be won, the life of our British coal-fields would be about 530 years, or, allowing for a wastage of 33 per cent., 350 years. It must be remembered, however, that the demand for fuel has, in the past, shown a continuous acceleration, and in his presidential address before the British Association in 1911, Sir Wm. Ramsay estimated that if the present rate of increase in the output were to be maintained, the life of these coal-fields would not exceed about 175 years.

In the United States, according to the U.S. Geological Survey the known coal-fields cover a total area of 310,300 sq. miles, to which may be added about 160,000 sq. miles of which little is known but which may contain workable seams, and about 32,000 sq. miles where the coal is at too great a depth to be worked under present conditions. The supply of coal before mining began is estimated to have been 2,750,000 million tons, consisting of 18,750 million tons of anthracite, 1,485,000 million tons of bituminous coal, 580,000 million tons of semi-bituminous coal, and 666,250 million tons of lignite.

The total production of coal to the end of 1911 has amounted to an aggregate of 7800 million tons, and this represents, including wastage in mining, an exhaustion of the beds equal to 12,650 million tons or somewhat less than 0·5 per cent. of the original supply, leaving as yet available some 2,740,000 million tons.

Regarding other parts of the globe, there is much greater uncertainty as to the magnitude of the available deposits. China is known to have deposits approximating 76,000 sq. miles in superficial area. No reasonably close estimate of the amount of these deposits is possible, but assuming them to yield the same quantity per sq. mile as is estimated for the coal-fields of the United States, this gives about 600,000 million tons. For other parts of the world, the most recent available estimates are as follow :

Germany	143,000 million tons.		
Austria-Hungary ...	17,000	,,	,,
France	17,000	,,	,,
Belgium	16,000	,,	,,
Russia, not less than	20,000	,,	,,
Canada ,, ,, ,,	100,000	,,	,,
India	10,000	,,	,,

Adding to these

Great Britain	...	145,000 million tons	
United States	...	2,740,000	,, ,,
China	... ...	600,000	,, ,,

and making an allowance of 192,000 million tons for other (unspecified) countries, we get a grand total of 4,000,000 million tons.

Huge as is this total, the aggregate output of the world's coal-fields is on a commensurate scale. The following approximate figures, referring mainly to the years 1905 and 1910, are taken from bulletins issued by the U.S. Geological Survey.

	1905		1910	
United States ...	350·7		448·0	million tons.
Great Britain ...	236·0		264·4	,, ,,
Germany	133·6		158·6	,, ,,
Austria-Hungary ...	40·7	(1909)	48·8	,, ,,
France	36·0		38·0	,, ,,
Belgium	21·8		23·5	,, ,,
Russia and Finland	17·1		22·3	,, ,,
Japan	11·9	(1909)	14·7	,, ,,
Canada	7·8		11·4	,, ,,
China	7·0	(1909)	11·8	,, ,,
India	8·4	(1909)	11·9	,, ,,
Australia	7·5	(1909)	7·1	,, ,,
Other Countries ...	15·2		21·2	,, ,,
Total	893·7		1081·7	,, ,,

For 1911, the output in the United Kingdom was

271·9 million tons; in the United States, 443·0 million tons; and in Germany, 158·2 million tons.

Huge as is this present rate of output, it will appear comparatively insignificant in the future if a corresponding rate of increase is maintained. At the present rate of consumption, the world's deposits, on the foregoing estimates, and allowing for 33 per cent. of wastage, would last some 2500 years; whereas should the consumption keep on increasing at the same arithmetical rate as during the past five years, they would be exhausted within 350 years.

Moreover, long before the fields are actually depleted, deep mining and heavy transportation charges must render their output comparatively costly in most of the present centres of industrial activity, and it is certain that from the present time the price of coal will rise steadily until it reaches a point at which other sources of energy, at present more costly and less convenient, will be able to compete on level terms. Even now anthracite has become too costly for use in many industrial processes, and its price is rising at a rate which promises to place it outside the class of ordinary fuels within the lifetime of the present generation.

Oil Fuel. Recent developments in the use of petroleum as a fuel for steam boilers and internal combustion engines, have led to the exploitation of oil-fields in many parts of the world. Deposits of oil-bearing strata, or of bituminous shale from which fuel-oil may be obtained by distillation, are widely distributed over the surface of the globe, as indicated in Fig. 1. By far the greater portion of the oil is at present obtained from wells drilled into petroliferous strata. Such strata occur in rocks of all geological ages from the Silurian upwards, but the most productive areas are the Palaeozoic in North America, and the Tertiary (Miocene) in the Caucasus. At the present time America, Russia, and Galicia are the largest producers, the former country yielding some 63 per cent. of the total output.

The principal petroleum deposits of the world are usually intimately associated with the principal mountain chains. In America, oil is found in Alaska and in a belt extending from Quebec to Tennessee; in California; from Nebraska along the Missouri river, and south to the Gulf of Mexico; in Cuba, Trinidad, and the northern portions of South America and southward as far as Peru and the Argentine.

In Europe, in addition to the famous oil-fields in the neighbourhood of the Caucasus and of the Black Sea, deposits are worked on the northern slopes of the Carpathians in Galicia, in Roumania, Transylvania, and Hungary. A small field is also worked in the neighbourhood of Hanover, while less important deposits are found in Sicily. Outside Europe and America, deposits—in some cases very extensive

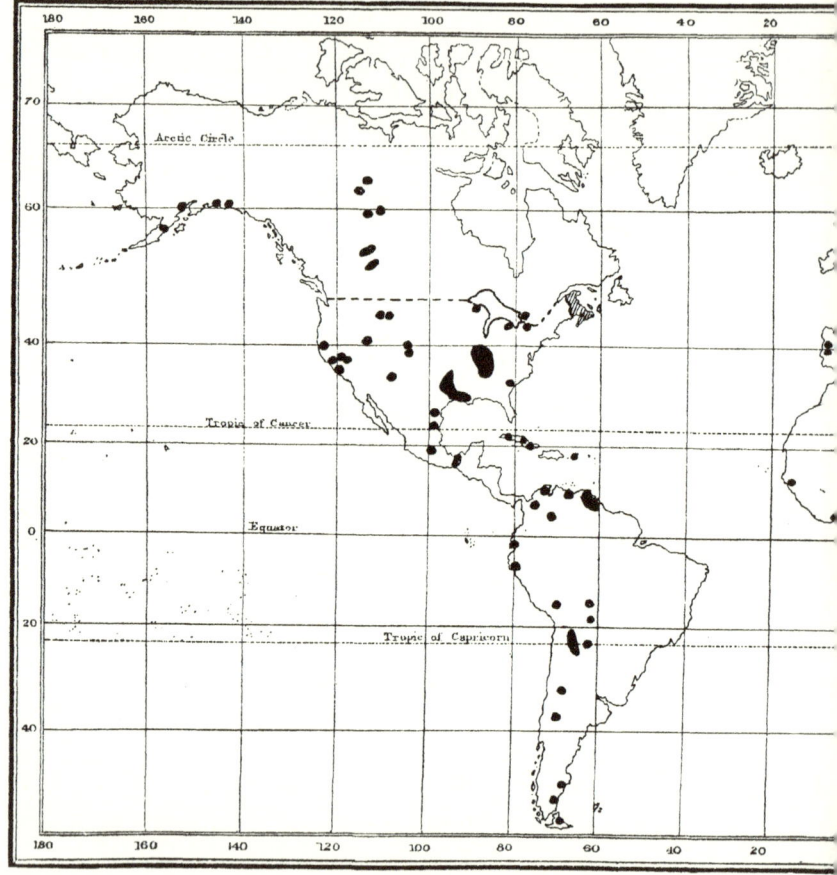

Fig. 1. The Principal Petroleum and Oil-Shale Deposits of the World.

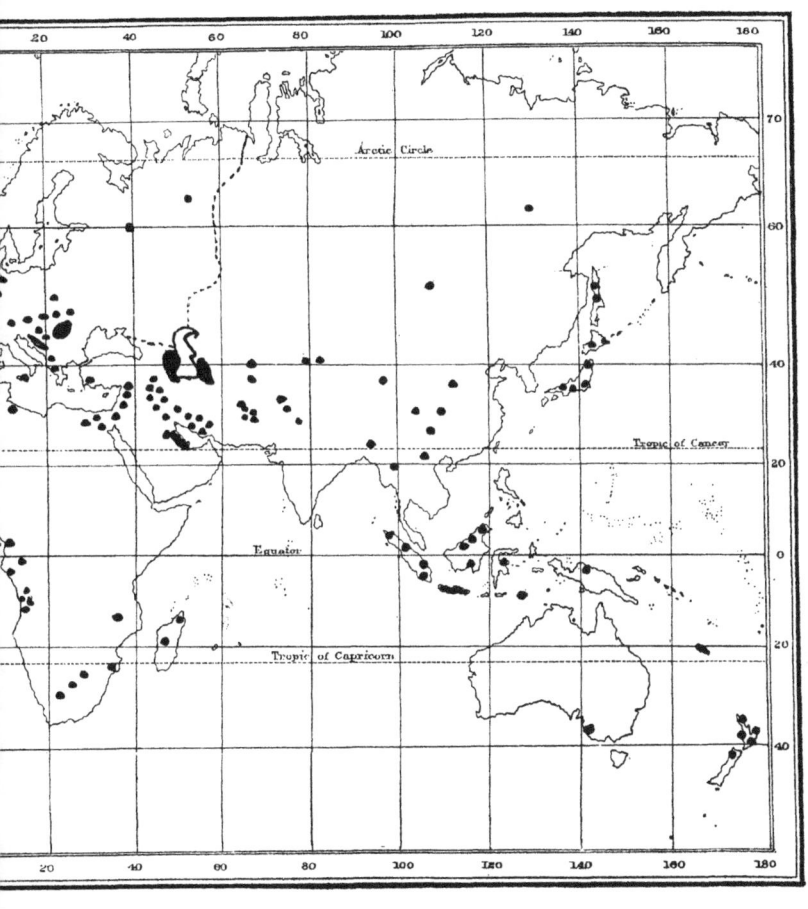

—are being opened up in Persia, Burmah, Borneo, Australia and New Zealand, and the world's annual output of oil, which has grown from about 26 million tons in 1905 to 52 million tons in 1911, is likely to be largely augmented in the near future.

The present output, if it could be entirely utilized as fuel for steam raising, would, however, only be equivalent to something under 7 per cent. of the world's output of coal. If used in internal combustion engines it would be equivalent, as a source of power, to about 15 per cent. of the coal raised.

Moreover, only a fraction of the oil is available as a fuel. A large proportion is, and always will be, required for lubricating and illuminating purposes. As the crude oil comes from the well, it is a complex mixture of many hydrocarbons, which varies in its chemical and physical properties according to the district from which it is obtained. This mixture is refined by fractional distillation yielding in turn petroleum spirit, lamp oil, lubricating oil, and a heavy residue (Astatki) which forms a valuable fuel. A typical oil will give the following approximate yield:

	American (per cent.)	Russian (per cent.)
Petroleum spirit ...	16	6
Lamp oil	54	20
Lubricating oil ...	18	6
Residue	2	58
Loss	10	10

Thus, in the most favourable grades of oil, not more than some 65 per cent. is available for power purposes.

Various theories are maintained as to the process of formation of petroleum. One school of thought attributes it to the distillation of such fossilized vegetation as coal; a second, to the decomposition of organic animal remains; and a third, to the action of water vapour under high pressures upon various metallic carbides. The oil-bearing strata in Europe belong chiefly to the Tertiary or later geological periods, and it is conceivable that in these strata or in those below them carboniferous deposits may exist and may be the sources of the oil, but in the United States and Canada the oil-bearing sands are found in the Devonian and Silurian formations which contain organic remains only in insignificant quantities. This and other cognate reasons give weight to the third of the above-mentioned theories. According to this, when in consequence of cooling, or for some other reason, the continuity of the earth's crust is disturbed, surface waters are enabled to make their way deep into the bowels of the earth and to reach the heated deposits of metallic carbides which in all probability are to be found in enormous masses at great depths. Under such circumstances, iron or whatever metal may be present forms an oxide with the oxygen of the water; hydrogen is set free and combines with the carbon of the carbide to form a volatile hydrocarbon ; and this, finding its way under pressure through any adjacent porous strata, is condensed wholly or in part as its temperature is lowered.

If this explanation be upheld, there is no reason why the manufacture of oil in Nature's laboratory should not proceed indefinitely. Indications, however, point to the fact that at the present time any such rate of manufacture is not commensurable with that of its depletion, and judging by the comparatively short life of many of the fields which have been exploited, the oil-bearing strata will be exhausted long before our coal-fields. The late Prof. Schaler [1] of Harvard University has indeed predicted that the supply of oil from wells sunk in the earth's crust will not sensibly outlast the present century, although this outlook is probably too pessimistic. Recent estimates by the U.S. Geological Survey give from 2000 to 4000 million tons as the probable resources of the United States and this, at the present rate of output, which, however, will probably be considerably accelerated in the immediate future, would, allowing for wastage, last for between 500 and 1000 years. While the output has steadily increased during the last 60 years, the increase has not been nearly in the same ratio as the number of wells. This is particularly the case in America, where many of the fields which, a few years ago, were among the largest producers, now rank very low in this respect. As a more permanent source of energy, the oil-shale deposits, such as are found in Scotland, France, Australia, Eastern Canada, and the United States are promising from a geological standpoint, although at present their output is relatively insignificant. The oil-shales of Scotland and of Australia are sufficiently rich in carboniferous matter to enable their product to compete in price, even now, with the output of oil wells. The Devonian shales of the United States are much poorer, and their oil cannot be produced at a cost much less than three times that now obtaining for such fuel. Still the magnitude of these deposits is so enormous that they are bound to play a large part in the fuel supply of the globe at no very distant time.

The Devonian black shale underlies a large portion of the Mississippi Valley, as well as of Pennsylvania and New York, and ranges in thickness from 25 feet to many hundreds of feet. The Utica shale, which is generally much thinner, underlies the northern Appalachian region. While such shales are generally better adapted to the production of gas than of oil, they furnish proportions of oil on distillation varying from 5 per cent. to 25 per cent. of their total weight. It is estimated that the State of Kentucky alone has 18,000 square miles of deposit, capable of giving 9 million tons of oil per sq. mile, or a total of

160,000 million tons, and it does not seem improbable that the oil-shale deposits of the States may ultimately provide as much fuel as the whole of their coal deposits. In the United Kingdom the only deposits of this nature are those of the Broxburn, Midlothian, and Linlithgow districts of Scotland. These have been successfully worked since about 1850, but are relatively insignificant in magnitude, their total yearly output of oil only amounting to about 300,000 tons. The shale is distilled in retorts at temperatures varying from 900° to 1300° F., steam being passed through the retort during the whole process. This enables the nitrogen present in the shale to be recovered in the form of ammonia, a ton of shale yielding from 180 to 270 lbs. of crude oil, and from 40 to 70 lbs. of ammonium sulphate. When refined, this crude oil yields approximately

Naptha   ...   .   6 per cent.
Lamp oil   32   ,,   ,,
Heavy fuel oils   26   ,,   ,,

As a fuel for steam raising, the heavy petroleum or shale oil residues possess many advantages over coal, and weight for weight will evaporate about 50 per cent. more water. Utilized as a fuel for use in an internal combustion engine, the oils, whether heavy or light, show a much greater economy. When used in this manner, 1 lb. of oil will generate as much mechanical energy as will 3·5 lbs. of coal in a good steam plant.

Natural Gas. The natural gas which is usually given off from borings in the vicinity of oil wells consists principally of methane, and is valuable either as a fuel for internal combustion engines or boilers, or for illuminating purposes. Extensive deposits have been drawn upon in the United States for many years, and, according to the records of the U.S. Geological Survey, about 500,000 million cubic feet were consumed in 1911. As a fuel for steam boilers, this is equivalent to some 10 million tons of coal, while if used as a gas engine fuel it becomes equivalent for power purposes to about 33 million tons of coal burned under a steam boiler.

In spite of the opening up of new deposits, the supply for 1911 shows a slight reduction as compared with that of 1910, which itself was practically the same as that of 1909. The gradual depletion of the older supplies in Ohio and Kansas is very apparent, and it is estimated that if the present rate of consumption is maintained many of the sources will be totally depleted within the next ten years. As a factor affecting the coal consumption of the country, this depletion of its gas supplies must largely counterbalance any gain which may be anticipated from the increasing use of oil fuel.

Peat. Of recent years much progress has been

made in the utilization of peat as a fuel for domestic
and industrial purposes. Extensive peat deposits
are found in various localities throughout Europe
and on the American Continent. The areas of the
European deposits are estimated to be approximately
as follow[2] :

		sq. miles
Russia and Finland		175,000
Scandinavia and Denmark ...		26,700
Germany		10,000
Ireland		1,800
		213,500

As taken from the bog, peat contains between
80 and 90 per cent. of water, and, on the average,
about 12·5 per cent. of dry substance. The weight
of dry peat in a bog one square mile in area and
10 feet deep is approximately 1,000,000 tons, and this
is equivalent for industrial purposes to some 500,000
tons of coal. Taking 10 feet as the average workable
depth, the bogs of Ireland are equivalent to 900
million tons of coal, and the total deposits of Europe
to some 100,000 million tons, or a little over one-
fourth of the estimated coal deposits of Europe
(including the United Kingdom).

The chief drawbacks to the use of peat on a large
scale arise from the high percentage and variable
proportion of moisture which it contains, and also
from the bulkiness of the material, which renders

transport costly.   Before being used as a fuel the greater portion of the moisture must be removed, and the difficulty is to find some economical method of doing this.   Artificial heating on a large scale is out of the question, and air-drying is the only practical alternative.   Even when air-dried the material contains some 25 per cent. of water, while in climates as humid as those in which the bogs are usually found the attainment of this degree of comparative dryness is a matter of some difficulty.   Such peat has a heating value about 40 per cent. of that of coal.

Much experimental work has been done of recent years, particularly in Germany and in Canada, with a view to modifying the ordinary gas producer for the use of peat as a fuel and to reducing the water contents of the material by mechanical processes.

In the former connection it has become possible to utilize peat containing as much as 50 per cent. of moisture, a degree which may be very easily attained by simple air-drying ; while, by heating the material under pressure to about 300° Fahr. and pressing to remove the moisture mechanically, it has been found possible to mould it into briquettes containing only about six per cent. of moisture, and having a calorific value of between 9000 and 10,000 B.TH.U.[5] per lb., or between 60 and 70 per cent. of that of coal. In spite of the necessity for, and the cost of such preparatory processes, the present consumption of

peat for industrial purposes is large in suitable localities. In the Moscow-Vladimir district of Russia the
consumption in 1909 was 500,000 tons. In this district the average distance between bog and mill is
about 7·5 miles, and the cost per ton at the mill yard
averages 9s. 4d., which would be equivalent to coal
at about 23s. per ton.

In districts more remote the cost of transport
would tell against the cost of peat as against that
of coal, and in spite of the magnitude of the deposits
they cannot be expected to relieve the demand for
coal and oil to any appreciable extent for many years
to come. Still, as the cost of these latter fuels increases, the peat deposits will doubtless be drawn
upon to an increasing extent.

Other sources of energy. In any case, sooner or
later it will become necessary to utilize some other
energy than that of the fossil fuels to do a fair share,
if not the major part, of the work of the civilized
world, and also to provide the domestic heating without which existence, on the present standard, would
be impossible during a large portion of each year in
the majority of our present centres of industrial
activity.

It remains, then, to consider what other sources
of energy are available for such purposes.

In this connection the following extract from a
paper by Sir Oliver Lodge[3] is suggestive of immense

possibilities. Speaking of the energy locked up in the aether, he says :

"As far as our present knowledge goes, an amount of energy equal to the output of a 1,000,000 kilowatt power station for 30,000,000 years exists, at present inaccessibly, in every cubic millimetre of space."

Fortunately it is impossible for us to liberate this energy at will, or the immediate results might be almost inconceivably disastrous. Still, as the behaviour of radium shows, the energy of molecular combination does, under some circumstances, undergo gradual transformation and degradation. Sir Wm. Ramsay[4] has estimated that the energy liberated by a pound of radium during its disintegration is 460,000 times as great as that of a pound of coal, or is equivalent to that utilized in a steam engine by the combustion of about 650 tons of fuel. Actually, radium takes about 1760 years for its natural disintegration ; but if it were possible to hasten this process until it took place at the rate of one pound in 24 hours, the energy liberated would be as great as that developed by an engine of 40,000 horse-power, and the energy of six pounds of radium would suffice to propel a vessel of the size of the *Olympic* across the Atlantic at a speed of 21 knots.

As the total amount of radium existing in the uranium ores of the world is estimated to be something under 600 pounds, it is evident that, so far as our present knowledge goes, not much is to be anticipated in this particular direction. But, apart from radium, it is probable that certain elements, which have hitherto been considered as permanent, are capable of being broken down with evolution of energy; and if any means could be devised for accelerating their almost inconceivably slow rate of change in such a way as to render their stores of energy easily and cheaply available, the energy problem would require no further solution.

Putting these fascinating possibilities aside, however, as being for the present even beyond the scope of rational speculation, it appears that the remaining sources of energy available to man may be roughly classified as follows :

(1)   Energy leaving the sun and reaching the earth as radiant heat.

(2)   Energy of fuels other than coal, oil, and peat, such as vegetable fibres and oils, timber, and alcohol, of which a regular and permanent supply may be ensured so long as present seasonal and climatic conditions continue.

(3)   Heat energy stored in the earth in virtue of the high temperature of its interior.

(4)   The energy of rivers and waterfalls and of all elevated stores of water.

(5)   The kinetic energy possessed by the earth in virtue of its diurnal rotation about its own axis.

Advantage may be taken of this by utilizing the rise
and fall of the tides as a source of energy.

(6) The energy of the waves of the sea.

(7) The energy of the winds.

With the exception of the earth's kinetic energy
and of its internal heat, all these are directly or
indirectly due mainly to radiant energy leaving the
sun.   The synthesis of carbon dioxide and water
to form the complex carbohydrates of growing vegetation is directly dependent on such energy ; the
energy of all elevated water-supplies, produced as
it is by evaporation of water from lower levels, is
mainly due to the same cause, and in a very minor
degree to heat transmitted by conduction from the
interior of the earth ; while the energy of the winds,
which are almost entirely caused by differences of
temperature and humidity, and of wave-motion,
which is directly due to the action of wind, may also
be attributed to the same source.

The following chapters are devoted to a more
detailed examination of these supplies of energy ; of
their possibilities as factors in any general power
scheme ; and of the possibilities of conserving our
present fossil fuels by improvements in our methods
of utilization.

CHAPTER II

THE TRANSFORMATION OF ENERGY

WHILE the energy of certain of the great natural sources, as for example the winds, tides, and waterfalls, is in a form directly available for mechanical purposes, the energy from other sources is not so directly available.

A piece of coal, for example, during its chemical combination with the oxygen of the air in combustion, liberates energy in the form of heat; a piece of zinc forming the cathode of an electric cell evolves energy in the form of an electric current during its chemical combination with the acid of the cell. If these sources of energy are to be utilized to drive some machine, it is necessary first of all to convert the heat or electric energy into mechanical energy.

On the other hand, mechanical energy, such as may be directly available at some hydraulic power installation, often has to be converted into electrical energy, in which form it is most easily transmitted through long distances, and a portion of this may afterwards require to be transformed into mechanical energy or into heat energy in some form of electric heater.

Hence this treatise is largely concerned with the transformation of energy from one form to another.

It is a well-known physical fact that heat and mechanical energy are mutually convertible, and that the transformation, when it occurs, always does so in a definite ratio. This fact was first discovered by Joule (1840), who, as a result of experiment, concluded that one British Thermal Unit (B. TH. U.) of heat is the equivalent of 772 foot-pounds of mechanical energy, and *vice versâ* [5]. More recent experiments indicate that Joule's original value is somewhat low, and 778 is now accepted as being more nearly the true value of the "mechanical equivalent" of heat.

Since one horse-power involves the expenditure of 33,000 foot-pounds of energy per minute, it follows that a supply of heat energy of magnitude 30,000 ÷ 778 or 42·42 B. TH. U. per minute is the heat equivalent of one horse-power.

The transformation from mechanical energy to heat takes place sensibly without loss at all temperatures. Thus, if an engine developing 100 H.P. at its crank shaft drives a drum on which is mounted a friction brake by which the power is absorbed, the heat given to the drum and brake will be 4242 B.TH.U. per minute, and this is sensibly true whether the drum be kept cool by means of a stream of water or whether, in the absence of cooling arrangements, it be allowed to get red or white hot.

On the other hand, owing, as will be seen, to the fact that the earth is at its present temperature, the reverse transformation from heat to mechanical energy cannot, under normal conditions, be carried out with any high degree of efficiency.

It may readily be shown that a perfect heat engine, *i.e.* one free from all radiation, conduction, leakage, and frictional losses, working on the most efficient of cycles, receiving its heat at an absolute[6] temperature T_1 and rejecting it in the exhaust at a lower temperature T_2, could only convert a fraction $\dfrac{T_1 - T_2}{T_1}$ of this into mechanical energy. For a given value of T_1 the efficiency of conversion increases as T_2 is diminished, and would be unity, *i.e.* the transformation would be complete if, and only if, T_2 were zero. The lowest value which T_2 can possibly have in any heat engine is, however, governed, for all practical purposes, by the normal temperature of its surroundings.   For example, in a condensing steam engine the working fluid cannot be cooled below the temperature of the available condensing water, and this determines the possible lower limit of T_2.   Evidently this cannot be lower than the mean natural temperature obtaining in the vicinity of the motor, which in this country has a value of about 60° F. or 521° absolute.   It is true that the condensing water might be cooled in a refrigerator before use, and this would enable the temperature and pressure of the exhaust steam to be reduced with a consequent increase in the work done by the engine. But the energy necessary to operate the refrigerator would be greater than the additional energy developed in the engine, so that the compound process would result in a nett loss.

It follows that if heat energy be supplied at say 700° absolute, not more than $\dfrac{700 - 521}{700}$ or 25 per cent. could possibly be transformed into mechanical energy, and even with an ideally perfect engine 75 per cent. would be rejected to the surroundings at a temperature of 521° absolute. If, however, the initial temperature were 7000° absolute instead of 700°, the proportion which might be transformed rises at once to $\dfrac{7000 - 521}{7000}$ or almost 93 per cent. This example illustrates with sufficient clearness the immensely greater possibilities, as regards transformation into mechanical work, of high-temperature as against low-temperature heat.

In this respect a dweller, say on Mars, with a mean annual temperature of about 400° absolute would be much more favourably circumstanced. Given heat at a temperature of 700° absolute, his same perfect engine would convert $\dfrac{700 - 400}{700}$ or 43 per cent. into useful work as against 25 per cent. on earth, while if his initial temperature were

$7000°$ absolute, the proportion would become
$\dfrac{7000-400}{7000} = 94$ per cent. as against about 93 per
cent. on earth.

The conversion of mechanical into electrical energy
in the dynamo-electric machine, or of electrical into
mechanical energy in the electro-motor, is an extremely efficient process. In either case some part
of the energy is utilized to overcome frictional and
eddy current resistances and is ultimately transformed into heat. This proportion is, however, small,
and in a modern machine the efficiency of conversion
in either direction is in the neighbourhood of 90 per
cent. Electrical energy may be converted into heat
by passing the current through a circuit in which
it has to overcome a comparatively large electrical
resistance, and so long as the resulting temperature
is lower than that of incandescence practically the
whole of the energy is transformed into heat.

The reverse process of converting heat directly
into electrical energy may be accomplished by the
agency of a thermo-electric cell, but the efficiency of
such a process is very low, and it is, so far, impracticable of accomplishment on any large scale.

The transformation of the energy of chemical
combination into heat or into electrical energy is
very efficient, and in the ordinary boiler furnace
or primary cell well over 90 per cent. of the energy

of oxidation of the fuel, or of the zinc plate as the case may be, is rendered available for further use. As the conversion of electrical to mechanical energy is also very efficient, the primary cell, in conjunction with some form of electro-motor, enables the energy of chemical combination to be converted into mechanical energy with comparatively little loss. Unfortunately, however, the cost of the materials suitable for use in a battery renders the method prohibitive on any large scale.

CHAPTER III

THE UTILIZATION OF THE FOSSIL FUELS

As long as the fossil fuels form the cheapest and most economical sources of energy for heating and power purposes, so long will their depletion continue to the neglect of other and less convenient sources. Nor can this or any succeeding generation be expected to conserve such resources for the benefit of posterity by a deliberate limitation of its industrial activities, or by any extensive utilization of other sources so long as these are less convenient or more costly. The elimination of all avoidable waste by the adoption of more efficient methods of utilization is, however, a different matter.   In so far as such methods may be adopted without accompanying disadvantages, there can be as little doubt that each generation is morally bound to do its utmost, as that many of our present methods leave an enormous scope for improvement in the way of fuel economy.

It is difficult to estimate with any pretensions to accuracy what proportion of the world's output of coal is used for various purposes. The following table, based on figures compiled by the Royal Commission on Coal Supplies, affords a rough idea of the quantities used for different purposes in the United Kingdom in 1910.

Power Generation.	Million tons
Factories	45
Mercantile and Naval Marine	13
Collieries	12
Railways	13
	83
Industries.	
Blast furnaces, steel and metallurgical processes	32
Chemical works, potteries, glass works, etc. ...	6
Gas works	20
	58
Domestic purposes	39
Total	180

As a fair proportion of the gas output of a gas-works is used in gas engines it appears from these figures that some 50 per cent. of the coal is used for the generation of power; 22 per cent. for domestic purposes; and the remainder for general industrial purposes. If these proportions be adopted as approximately true for the world's output, it follows that the coal used for power generation at present approximates to 600 million tons per annum.

Taking into consideration the whole of the steam plants in existence, old and new, it is estimated that on the average as much as 5 lbs. of coal are required per H.P. hour, while many of the older and smaller plants use at least four or five times this amount. Assuming an average of 5 lbs. per H.P. hour with 3000 working hours per annum, this means that the horse-power of the globe, dependent on coal as a fuel, is of the order of 100 millions.

The modern efficient steam plant, with the usual aids to economy in the way of steam superheater and economiser, consumes between 1·5 and 2·0 lbs. of coal per H.P. hour including all stand-by losses, and assuming it to be possible to bring down the average consumption from 5 lbs. to 2·5 lbs. by scrapping the worst of the plants, this alone would involve an annual saving of some 300 million tons, having a value at 10s. per ton of £150,000,000.

It is a question for argument to what extent such improvements should be left to the option of the individual. Every inefficient steam plant involves an indirect tax, not only on all present and future consumers of coal by the mere operation of the law of supply and demand, but also on the community at large.

It is true that the scrapping of an inefficient power plant and its replacement by one of modern design is a costly operation. Still such an alteration would usually pay for itself within a comparatively few years, and, in the average case, with a plant using 5 lbs. of coal per H.P. hour, less than 10 years should suffice to write off all charges. In many cases where such a change might reasonably be expected it is not made either through indifference or through conservatism ; in other cases doubtless because of difficulty in raising the requisite capital. In a community in which the interests of the whole body were paramount the logical outcome would be a State Conservation Department with powers to compel and if necessary subsidise such alterations as might be necessary; and while no public authority is invested with any such powers at present, signs are not wanting that such a department will form one of the most important adjuncts of the future State.

Domestic Heating. Large as is the proportion of heat wasted in power generation, it is exceeded by that involved in the methods of domestic heating usual in this country. It is estimated that something under 12 per cent. of the heat from the coal burnt in an open fireplace goes to warm the room, while the ordinary cooking range only utilizes some 5 per cent. of its heat supply. The systems, common in the United States and in some continental countries, of heating by means of steam, hot water, or hot air, from some central heater, or by means of slow combustion stoves, are infinitely better from a fuel efficiency point of view. Although the former methods are now fairly general in this country for the heating of large institutions, the central heater is still as unusual in our homes as is the open fireplace in the United States.

The open fireplace with its cheerful aspect and hygienic advantages has certainly much in its favour, but it is doubtful whether the strong prejudice of the average Englishman against its abolition is not mainly due to conservatism. The emigrant to the United States or Canada seldom expresses any strong desire to return to its use after a few years in his adopted country, although the winters in most parts of both countries are much more severe than our own.

It is true that the use of gas stoves and of electric radiators for heating purposes is increasing yearly, and that gas cookers for culinary operations have come into very general use. At present the wider adoption of such methods is largely a question of the cost of gas and of electric current. With coal gas at 2s. per 1000 cub. feet and electricity at 1d. per unit, gas is found to be about seven times the cheaper of the two in actual practice.

From the point of view of fuel economy indeed, electric heating is in most cases even worse than the open fireplace. If the electricity is generated by a steam plant using 2 lbs. of coal per H.P. hour, this is equivalent to an output of 1273 heat units per lb. of coal. Frictional losses in the steam engine, and frictional and electrical losses in the dynamo, along with losses in transmission, inevitably reduce the energy delivered to the radiator to something under 75 per cent. of this, so that not more than 950 B.TH.U. per lb. or a little over 6 per cent. of the heat value of the fuel is converted into heat at the radiator. With an internal combustion engine as the source of power, using producer gas generated from coal, this might be increased to some 12 per cent., but bearing in mind the cost of interest, depreciation, and attendance of the power plant, it is evident that the cost of electrical heating generated from fuel must always be excessive. With electricity generated on a large scale from some cheaply available source of hydraulic energy the problem becomes more hopeful, and in some recent installations of this type the cost of such heating is claimed to be little if any greater than that of heating by coal fuel even at the present price of the commodity.

The use of coal gas as a fuel for domestic heating

is more promising.   When heated in a retort, gas coal will give off coal gas having a total heating value about 18 per cent. of that of the coal.   In addition to this, gas coke having a heating value of some 60 per cent. of the coal, is left as a residue which may be used as a solid fuel, while the gas tar which is also produced may, under certain conditions, be utilized as an oil engine fuel.   The total loss of heat in the process of gasification does not therefore exceed some 20 per cent.   Experiments show that an efficient gas stove, provided with a sufficient length of piping between stove and flues to cool the products of combustion to about 250° F. before allowing them to escape from the room, will give the same heating effect with 4 cubic feet of coal gas as will the ordinary open grate with $1\frac{3}{4}$ lbs. of coal, or the modern type of well-grate with 1 lb. of coal.   With coal gas at 2s. per 1000 cubic feet and coal at 18s. per ton the costs of the two methods of heating are identical if the coal is burned in a well-grate, while if the more usual type of grate is used the cost of heating by gas is little more than one-half that of coal heating.

The price of gas is often needlessly inflated by the enforcement of unnecessarily high illuminating standards, but where it is required only for heating or power purposes the whole of the combustible matter of the fuel may be gasified in some form of producer, and up to 80 per cent. of its heat value may then be obtained in the resultant gas. As, with a well-designed heating stove, some 85 per cent. of this heat may be usefully utilized, the overall efficiency of such a process is of the order of 68 per cent.

Using a central steam or hot water system with a good boiler and properly covered supply pipes it is possible to utilize some 66 per cent. of the total heat value of the coal fuel for actual heating, or practically the same as with gas heating. In the average small house, burning say 12 tons of coal per year in open grates, the change to central heating would thus result in an annual saving of about 9 tons of coal, with an average value of about £10, a sum which, capitalised, would be more than sufficient to defray all the expense of any necessary alterations. At a moderate estimate, the general adoption of such a system of heating in this country alone would involve an annual saving in fuel of the order of 35 million tons—truly a weighty consideration.

Fuel for Power Purposes. When required for the generation of power any solid fuel may be used in either of two ways. It may be burned directly in the furnace of a boiler supplying steam to an engine or turbine, or may be gasified in a gas producer and may then be burned in the cylinder of an internal combustion engine. While some liquid fuels may also be used in the furnace of a boiler, these are more generally suited for use in an internal combustion motor, and for such a purpose may generally be gasified without the expenditure of any otherwise useful heat.

A good steam boiler, if carefully fired, will convert up to 80 per cent. of the total heat of its fuel into energy in the form of high pressure steam, so that no great improvement is possible in this direction.

Unfortunately, however, the temperature at which this steam is formed is so comparatively low that its availability for transference into mechanical energy is not very great [7] ; and with the highest pressures adopted in practice the most perfect engine working on any practicable cycle could not transfer more than about 30 per cent. of the heat into useful work. Losses inseparable from any engine or turbine using steam as its motive fluid reduce this proportion still further, and in the most efficient of modern steam plants only about 20 per cent. of the energy in the steam, or 16 per cent. of the energy of the fuel, is usefully utilized. Nor is there any possibility of appreciably increasing this proportion by improvements in design and construction. Something might certainly be done in the way of increasing working pressures and temperatures, but with fuel at its present price the additional cost of the necessarily heavier construction, along with the increased cost of maintenance, would more than counterbalance the possible gain in fuel economy.

By burning a fuel in the cylinder of an internal combustion engine, many of the losses inherent in the steam plant, such as radiation from the boiler and supply pipes, and condensation in the cylinder, are eliminated, while owing to the much higher temperature of the working fluid—some 2700° F. as against some 400° F.—its possibilities as regards transformation of its heat into mechanical energy are greatly increased. In a modern gas engine up to about 33 per cent. of the heat energy of the gaseous fuel is converted into work at the crank-shaft or twice as much as in the best steam plant.

Gas fuel and gas producers. During the distillation of 1 ton of gas coal to form coal gas, approximately 10,000 cubic feet of gas having a heating value of 600 B.TH.U. per cubic foot, along with 1500 lbs. of coke and 110 lbs. of gas tar, are produced. The heating value of the gas is thus about 18 per cent. and that of the coke about 58 per cent. of that of the coal used. When used as a gas engine fuel, about 13 cubic feet of such gas is required per H.P. hour, so that the gas produced from one ton of coal would be equivalent to 770 B.H.P. hours, or at the rate of 3 lbs. of coal per H.P. hour. This is practically the same as the consumption of a moderately good steam plant, and as the whole of the coke is available for power production the use of a fuel in this way is much

better—from a conservation point of view—than its
use in a steam plant.

Coal gas is, however, primarily intended for
illuminating purposes, and the consequent purification
and enrichment which are necessary to render it
suitable for such work add to its cost to an extent
which renders it unduly expensive for power use on
any large scale.

In some few districts coal gas is sold for as little as
1s. per 1000 cubic feet, but in general its cost is at
least 2s. per 1000 feet and at this rate it costs about
·32d. per H.P. hour, as against an average fuel cost of
about ·16d. per H.P. hour in the case of a steam plant.
At 1s. per 1000 feet the costs of fuel in the two cases
are practically identical.

Where a gas is required only for fuel or power
purposes the necessity for any high degree of
purification vanishes, while for engine work the
presence of the heavy hydrocarbons which render
coal gas valuable as an illuminant is rather a dis-
advantage than otherwise, as it limits the degree of
compression which can be obtained in the cylinder
without risk of preignition.   For such purposes a
much cheaper, if poorer, gas can be made by passing
air, steam, or a mixture of air and steam through a
column of incandescent fuel contained in the enclosed
furnace of a " producer."

When air alone is used the carbon in the lower

portions of the fuel bed burns to carbon dioxide, which is reduced to carbon monoxide by contact with the heated carbon in the upper portion of the producer, and a gas is given off which is essentially a mixture of carbon monoxide and nitrogen with the addition of a little hydrogen (from the fuel) and carbon dioxide.

A typical analysis of the gas from a producer working on this system and using coke as fuel gives

CO	...	...	...	32·6 $^\circ/_\circ$
H	...	...	...	1·0 $^\circ/_\circ$
CO_2	...	...	...	1·4 $^\circ/_\circ$
N	...	...	...	65·0 $^\circ/_\circ$

The heating value of such a gas is low—about 115 B.TH.U. per cubic foot—but owing to its high percentage of carbon monoxide and the comparatively small volume of air necessary for its combustion (about 1 cubic foot per cubic foot of gas) it forms a very suitable gas for gas engine work.

Owing to the amount of heat evolved during the first combustion of the fuel to carbon dioxide in this process the temperature of the outgoing gases is high—usually about 850° F. Where the gas can be used hot, as in the heating of gas retorts, for glass melting, etc., this is no drawback and the process is very efficient. Where, however, as in gas engine work, it is necessary to cool it down before use this

involves a comparatively large loss of heat, and under
such circumstances only between 60 and 70 per cent.
of the heat of the fuel is available in the cooled
gas.   Also the high furnace temperatures which are
produced by this method of working give rise to
clinker troubles with most fuels.

The admission of steam or water vapour along
with the air supply to the furnace serves two useful
purposes.   In the first place it tends to keep down
the temperature of the furnace and hence to reduce
clinker troubles.   Secondly, the steam combines with
incandescent carbon to form hydrogen and carbon
monoxide, both of which go to enrich the gas.   The
interaction of water vapour and carbon is an endo-
thermic process, that is it absorbs heat, and the heat
of combustion of the resultant gases is greater than
that of the carbon involved, by the amount of heat
absorbed.   When this action takes place in a
producer it absorbs a portion of the heat evolved in
the combustion of carbon to carbon monoxide, less
heat is wasted in heating up the outgoing gases and
in radiation, and the overall efficiency of the process
is appreciably increased.   The gas produced contains
more hydrogen and carbon dioxide, but less carbon
monoxide than with a dry air blast.   Its composition
depends largely upon the amount of water vapour
admitted with the air, and also on the composition of
the fuel.   The best results are obtained when about

·5 lb. of vapour is admitted per lb. of fuel, and under these circumstances the heat of combustion of the gas, even when cooled, is about 80 per cent. of that of the fuel. With more vapour than this a portion passes through the furnace without being decomposed, carrying with it its latent heat of vaporization and so reducing the efficiency of working. When working under the best conditions the gas from such a producer, using anthracite as a fuel, has a composition approximating to

CO	...	...	25 %
H	...	...	18 %
CH_4 (methane) ...	...	1 %
CO_2	...	...	6 %
N	...	...	50 %

with a heat value of about 143 B.TH.U. per cubic foot. Such a gas forms a fuel which is almost ideal for gas engine work, containing as it does enough hydrogen to render it easily ignited in the cylinder, but not too much to enable fairly high compression pressures to be used without danger of preignition.

By using a large excess of water vapour in the air supply, the furnace temperatures may be considerably reduced, and this fact is taken advantage of in the "Mond" producer to enable the ammonia which is produced during combustion of the bituminous fuel used, to be recovered. In other plants ammonia is decomposed at the high temperatures attained, and to prevent this about 2·5 lbs. of water vapour are admitted to the furnace per lb. of coal.   Only about ·5 lb. of this is decomposed, and to recover as much as possible of the latent heat of the excess vapour, the outgoing hot gases and steam are led through a series of pipes in a tubular regenerator, while the fresh air and steam supply pass over the outside of the pipes on their way to the furnace and take up part of the heat. By this means the heat efficiency of the process is raised to between 75 and 80 per cent.   The gases are then passed through a washer in which they are cooled and the tar is removed, after which they are led through a tower in which they meet a current of acid liquor containing about 4 per cent. of free sulphuric acid.   Here, ammonia is converted into ammonium sulphate, of which from 70 to 90 lbs. are produced per ton of fuel.

The gas produced has a composition approximating to

H	...	...	...	$25\,\%$
CH_4	...	...	...	$2\,\%$
CO	...	...	...	$13\,\%$
CO_2	...	...	...	$13\,\%$
N	...	...	...	$47\,\%$

with a calorific value of about 152 B.TH.U. per cubic foot.

The value of the ammonium sulphate produced as a by-product is practically the same as that of the fuel used (at present prices the value of the sulphate produced per ton of fuel is about 10s. 6d.), but owing to the high first cost of the recovery plant, the process is not commercially economical for an installation of less than about 4000 H.P.

For many metallurgical processes the comparatively large proportion of inert gas—mainly nitrogen—found in the ordinary producer gases, and the consequently comparatively low temperatures obtained during combustion, form a drawback. This may be overcome by blowing steam alone through the layer of incandescent fuel in the furnace, in which case the gas given off—known as water gas—has a composition approximating to

H	...	...	...	50 °/$_{\circ}$
CH_4	...	...	...	1 °/$_{\circ}$
CO	...	...	...	40 °/$_{\circ}$
CO_2	...	...	...	4 °/$_{\circ}$
N	...	...	...	5 °/$_{\circ}$

with a value of about 260 B.TH.U. per cubic foot. Such a gas, though not so suitable for use as a gas engine fuel as the leaner producer gases, is particularly well adapted for manufacturing processes such as the welding of boiler plates and tubes, the heating of open hearth steel furnaces, etc., where an intense local heat is required.

The manufacture of water gas is an intermittent process. In order to raise the bed of fuel to a high temperature a strong air blast is passed through for about 1 minute. The air is then cut off and steam is blown through for from 7 to 10 minutes. This reduces the temperature of the fuel, after which it is again air-blasted, and so on. During this portion of the process if the blast is sufficiently strong the resultant gas is mainly carbon dioxide and nitrogen, and in order to conserve the heat evolved in combustion, these gases are led through a regeneration chamber filled with a chequer of brickwork which is raised to a high temperature. The ingoing steam is led through this chamber on its way to the furnace, absorbing its heat and becoming superheated in the process, and by this means the overall efficiency of the plant may be increased to between 70 and 75 per cent.

Suction and Pressure Producers. In a gas producer the air or mixture of air and steam necessary for operation may be forced through the furnace by the action of a blower or of a steam injector, or, if the producer is supplying gas to the cylinder of a gas engine, may be drawn through the furnace by the pumping action of the piston during the suction stroke. In the former case the plant is termed a "Pressure" producer; in the latter a "Suction" producer.

In the suction producer the engine itself automatically regulates its own supply of gas to suit the demand for power, and in view of the simplicity and ease of working of such a combined plant its use, particularly for comparatively small powers, has become very common. Fig. 2 shows, diagrammatically, the general arrangement of such a plant.

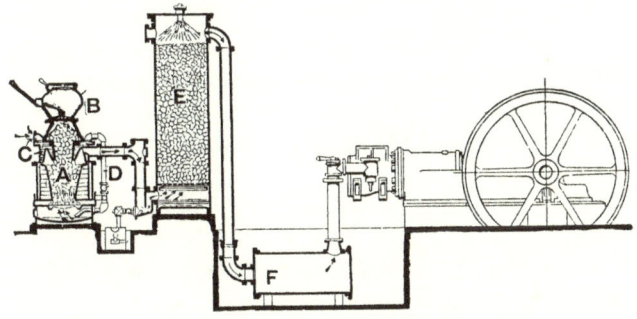

Fig. 2. General arrangement of Suction Gas-Producer.

Here A is the generator supplied with fuel through the hopper B. Air enters the vaporizer C which contains water heated by the hot gases leaving the generator, and in its passage over the surface of this water saturates itself with its vapour. The mixture of air and vapour is led through the pipe D to the under side of the grate, and is then drawn through the furnace, emerging at the top in the form of producer gas. This is passed through the scrubber *E* on its way to the expansion box *F*, from which it is drawn on each suction stroke of the engine. The scrubber consists of a wrought-iron chamber, loosely packed with coke over which water is sprinkled, and serves to cool the gas to atmospheric temperature and to remove the dust and tar which it carries in suspension.

Unfortunately the present types of suction producer can, in general, only be operated successfully on non-bituminous fuels, such as coke and anthracite. With bituminous fuels much tar is produced, and in order to prevent sticking of the valves of the engine, it is essential that this be thoroughly removed between the generator and the engine. In the pressure producer this may readily be accomplished by increasing the number of scrubbers and by introducing one or more gas washers between producer and engine, the additional resistance which these involve being overcome readily by a slight increase in the pressure at the blower. In the suction producer, however, the introduction of an elaborate cleansing apparatus increases the necessary suction at the engine and thus diminishes the weight of gas which can be drawn into the cylinder per stroke to an extent which seriously reduces the power which can be developed with a given size of engine, and thus increases the first cost of the plant in a double degree. Moreover, owing to the fact that the bituminous fuels are more prone to clinker troubles, and usually contain more ash than non-bituminous coal, more frequent cleaning of the fire and grate is necessary, and this is more easily performed without deranging the operation of the plant where the latter is of the pressure type.

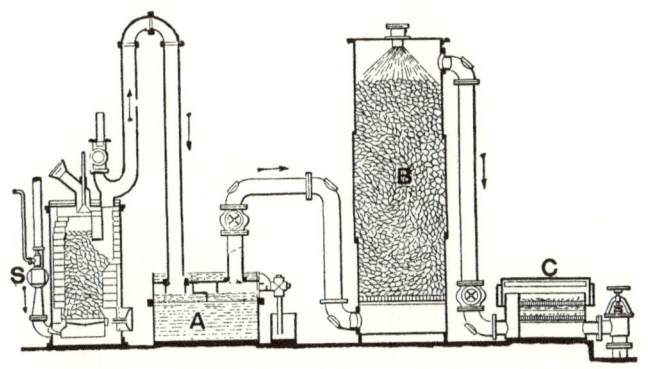

Fig. 3.   General arrangement of Pressure Gas-Producer.

Fig. 3 shows the general arrangement of a pressure producer for bituminous fuel.

This differs essentially from the suction producer in that a mixture of air and steam is blown through the furnace by means of the steam injector *S*, which is supplied from an auxiliary steam boiler working at a pressure of about 60 lbs. per square inch. After leaving the generator the gas is passed through water in the chamber A, in which part of its tar is removed. It is then taken through the coke scrubber B, and afterwards through a sawdust scrubber C in which the final traces of tar are removed. On leaving this scrubber it is led into a gas holder, from which the various engines or furnaces derive their supply.

In the pressure producer any type of non-caking fuel may be used, and considerable progress has been made in adapting it to such fuels as peat, sawdust, and waste vegetable fibre. As a fuel, peat is particularly rich in nitrogen, and ammonia recovery is therefore comparatively more advantageous. Its use as a producer fuel in conjunction with recovery apparatus of the Mond type is increasing to a large extent on the Continent of Europe. Experiments[9] show that one ton of dry peat with about one-half the heat-value of bituminous coal will produce practically one-half as much producer gas as coal, but will yield some 120 lbs. of ammonium sulphate per ton as against 80 lbs. in the case of coal. Probably the cost of cutting and drying a ton of peat is as much as that of a ton of coal delivered at the same place, so that for power purposes coal is at present the cheaper fuel. Moreover, a peat producer requires to be much larger and therefore more costly for a given power.

Still, where a plentiful supply of good peat is available, there is no reason why it should not be advantageously employed in this way, and it is probable that this will prove an important development in the future application of the power gas producer.

Taking the heat efficiency of the average good producer as 80 per cent., it appears that the complete plant, consisting of producer and gas engine, will convert some 25 per cent. of the heat of its solid fuel into mechanical work, or twice as much as a good steam plant using the same fuel. Under present conditions, indeed, such a plant provides by far the most efficient means of utilizing the energy of a solid fuel.

In the ideal installation the producer, or battery of producers, would be situated at the pit mouth. The gas might then be distributed for considerable distances through pipe lines without any great loss, to be used in gas engines or for heating purposes where required, or, alternatively, might supply a battery of gas engines near at hand from which power would be transmitted electrically. The former method has been adopted in such a plant[8]—probably only the first of a large number—erected on the South Staffordshire coal-field. This plant consists of eight Mond producers ; has a capacity of some 200 tons of coal per day ; and distributes its gas under a pressure of about 7 lbs. per square inch to numerous small towns spread over an area of about 120 square miles. The average charge for this gas is 1·8*d*. per 1000 cubic feet.

In his presidential address to the British Association in 1911, Sir Wm. Ramsay suggested going even further than this, and proposed, by gasifying the fuel *in situ*—underground—to obviate the necessity for winning and hoisting the fuel. The practical difficulties in the way of such a scheme are of course very great, the chief of these being the difficulty of supplying the air necessary for partial combustion and of regulating its supply as required to every portion of the underground area. On the other hand, the potential advantages of the scheme are great. Not only would it enable the heavy costs of winning, hoisting, and transportation of the fuel to be eliminated, but seams too thin or of too poor quality to admit of being worked successfully under present conditions might be utilized as readily as the thicker and richer seams ; also the wastage which is involved by the necessity for propping the working galleries of the ordinary mine would be eliminated. As it is estimated that under present conditions of working not much more than one-half the coal available in the average field is ever raised to the surface, the importance of these latter considerations is evident. Facilities have been afforded for the trial of this method on a fairly large scale in one

English pit, and the results will be awaited with much interest.

The Internal Combustion Engine. Whatever type of gas is used in a given internal combustion engine, the power developed is approximately the same, owing to the fact that the richer gases require a greater admixture of air for perfect combustion. The resultant mixture of gas and air drawn into the cylinder is thus of sensibly the same heat-value per cubic foot in every case, coal gas enabling only about 10 per cent. more power to be developed in a given engine than ordinary producer gas. The thermal efficiency of the engine, *i.e.* the ratio of the heat turned into work to the heat of the gaseous fuel, depends mainly on the degree of compression of the working mixture at the beginning of the explosion stroke. The magnitude of this is limited by the necessity for avoiding preignition during compression, and with the compressions usual in the ordinary gas engine working on producer gas, efficiencies of between 28 and 32 per cent., measured on the brake horse-power, may be obtained.

Liquid fuels having a comparatively low temperature of vaporization, such as petrol or kerosene, may be vaporized in a carburettor, and the vapour thus produced, when mixed with air, may be used as the working fluid in an ordinary gas engine. All petrol motors and many small oil engines work on this principle. In a petrol motor, owing to the fact that the liquid vaporizes at ordinary atmospheric temperatures, no difficulty is experienced in producing the carburation of the ingoing air, whereas kerosene requires the application of external heat. This is usually abstracted from the hot exhaust gases, whose heat would otherwise be wasted, so that this necessity does not involve any reduction in the thermal efficiency of the plant. Owing to the large proportion of hydrocarbon in the explosive mixture in either case, the maximum limit of compression is considerably less than is the case with the engine using fuel gas, and the efficiency of the motor is thus only about 22 to 26 per cent.

To avoid this drawback, large modern engines using oil-fuel work on the Diesel principle. The course of operations of such an engine working on a four-stroke cycle is as follows:

(1) Air at atmospheric pressure and temperature is drawn into the cylinder.

(2) This air is compressed on the inward stroke of the engine to between 400 and 500 lbs. per square inch. In the process its temperature becomes raised to about 950° F., or far above the ignition-point of oil.

(3) At the end of the compression stroke, oil is sprayed into the cylinder by means of compressed air under a pressure of 700 to 800 lbs. per square

inch.   On mixing with the hot air in the clearance
space of the cylinder this ignites and burns.   During
the first portion of the stroke, while the oil is
being sprayed in, the piston is moving forward and
combustion takes place at approximately constant
pressure.   When about one-tenth of the stroke has
been completed, the oil supply is cut off, and there-
after the hot gases do work by expansion to the end
of the stroke.

(4)   The exhaust valve opens and the waste
gases are exhausted to atmosphere.

Many large modern engines have been designed
to work on a two-stroke cycle, in which case an air
blast is admitted to the cylinder at the end of the
power stroke.   This blows out the burnt gases.   The
air in the cylinder is then compressed on the inward
stroke, fuel is injected, and the next outward stroke
is again a power stroke.   By this means the power of
an engine of given dimensions is practically doubled.
The chief difference between the Diesel oil engine
and the ordinary gas engine lies in the fact that the
only substance in the cylinder during the compression
stroke is air, with possibly a little of the burnt gases
from the preceding cycle.   Preignition before the
end of this stroke becomes impossible, and much
higher compression pressures may therefore be
adopted than are possible when the working fluid
is itself compressed.   As a result of this the thermal

efficiency of such an engine is high, and usually amounts to about 35 per cent. on the brake horse-power, or over 40 per cent. on the indicated horse-power.

The Diesel engine is capable of working on a very wide range of fuels, and experiment shows that tar oils distilled from lignite and fat oils from animal or from vegetable sources, such as earthnut oil, castor oil, train oil, fish oils, etc., can be used quite as readily as mineral oils. By slightly modifying the plant it is also possible to utilize tar distilled from pit coals, and also vertical-oven, water gas, and oil gas tars, the only tars which cannot be used being those from horizontal or inclined gas, or coke, retorts. At the present time the petroleum residues form the most important fuel supply, but it is not impossible that in the remote future the oils from vegetable sources may occupy the same position as the petro-leum products do to-day. With an adequate supply of such oils, the internal combustion engine would be independent of the fossil fuels, and, in this direction, at least a partial solution of the energy problem of the future is likely to be found.

Relative Costs of Operation by Steam, Gas, and Oil. Taking average modern types of plant in good order, the weights of fuels used, and the fuel-costs per B.H.P. are roughly as follows:

Plant		Type of fuel	Cost per ton	Weight per B.H.P. hour (in lbs.)	Cost of fuel per B.H.P. hour (in pence)	Total cost per B.H.P. hour (in pence)
Steam		Bituminous coal	12/-	2·0	·13	·37
Gas engine with	Suction producer	Anthracite	25/-	·8 to 1·0	·11 to ·13	·36
		Coke	10/-	1·25 to 1·5	·07 to ·08	·31
	Pressure producer	Bituminous slack	10/-	1·0 to 1·25	·055 to ·07	·30
Diesel oil engine		Petroleum residues	£3	·45	·145	·35

The fuel-costs, of course, only form a small proportion of the total charges of a power plant, and when the interest on the capital cost of the plant, depreciation charges, and costs of lubricating oil and attendance are taken into account the relative costs of the various types of plant are changed somewhat.

Taking an electrical installation of 2000 B.H.P., the total cost of plant, buildings, and foundations, with all necessary auxiliaries, cables, and switchboard, is practically identical with each type of plant, amounting to about £18 per B.H.P. With a load factor of 52 per cent., the running cost, including wages, repairs, lubrication, and water, amounts to about ·13*d*. per B.H.P. hour for either a steam or gas installation and to about ·095*d*. for a Diesel installation, so that the total charges, including 15 per cent. for interest and depreciation of machinery, are roughly as given in the last column of the preceding table. While these figures are only very rough approximations, and will never be quite the same even for two identical plants, they do indicate that there is no economic reason why the average steam plant should not be replaced in the majority of cases by the gas or oil engine plant.   Even with up-to-date steam plants this would reduce the draft on our fossil fuel supplies by one-half, while if the average steam plant be taken as the standard, not much more than one-quarter of the fuel presently consumed for power purposes, would be necessary.

The production of electricity directly from coal. Were it possible to use carbon instead of zinc as the positive electrode of a voltaic cell, and to oxidise this carbon electrolytically, electrical power might be obtained directly from coal, instead of, as at present, through the chainwork of the boiler and steam engine, or gas producer and gas engine, and dynamo, and in such a case some 90 per cent. of its energy of combination might be usefully utilized instead of about 10 per cent. with the steam plant and 25 per cent. with the gas plant.

In view of the huge saving of fuel which this

would permit, many attempts have been made to solve the problem of the carbon cell, but with small success. The difficulties arise from the fact that, while carbon is a good conductor of electricity, it is electro-negative to all but a few elements, and thus the choice of a suitable negative for the cell is strictly limited ; also carbon is insoluble in all but a few solutions. One or two inventors have attained a partial measure of success in their endeavours, though in each case an external heat supply, with its accompanying fuel consumption, is necessary. Moreover, in proportion to the output of energy, such a plant is extremely cumbrous and costly. Still the development of such a cell on commercial lines must rank among the possibilities of the future, and although it would probably do little to reduce the cost of power, its influence in reducing the consumption of coal, and in conserving our resources, might be very appreciable.

CHAPTER IV

THE UTILIZATION OF SOLAR HEAT

THE amount of radiant energy received by the earth from the sun and manifested in the form of heat is, in the aggregate, enormous. At points between the equator and latitudes 45° north and south it amounts on an average to the equivalent of 8000 foot-pounds per minute for each square foot of area exposed to the perpendicular rays of the sun, and if this could be entirely converted into mechanical energy one horse-power might be developed for each four square feet of heating surface.

Furthermore, owing to the intensely high temperature of the sun's surface—some 6000° C.—this radiant energy would be in an ideal state for efficient conversion if only some means of direct transformation were practicable. Unfortunately, so far as our present knowledge goes this is impossible, and we have to be content to use it to heat water or some other volatile fluid such as sulphur dioxide or ether, and to use the vapour from this in some form of steam engine or turbine. In practice the temperature of the working fluid does not usually exceed some 100° C., and the degradation of the energy to this temperature means that its potentiality for transformation into mechanical energy is reduced to about one-fifth of its original value. Thermal and mechanical losses in the engine reduce this still further, with the result that in practice a minimum of at least 100 square feet of heating surface is necessary for the production of one horse-power. This is equivalent to 280,000 H.P. per square mile of surface.

The utilization of this radiant energy has been

attempted by several inventors with some measure of success.

Mouchat in France, and afterwards Ericsson (1868–1875) in the United States, used a large

Fig. 4. 10 Horse-Power Solar-Heat Plant on an Arizona ranch.

parabolic mirror, in form like an inverted umbrella (Fig. 4), to reflect and concentrate the sun's rays on a small boiler placed at its focus and in this way obtained 1 horse-power for slightly under 100 square

feet of surface.   In a somewhat similar and larger
plant tested at Pasadena (Cal.) in 1898, the reflector
consisted of a parabolic structure built up of plane
rectangular mirrors each arranged so as to throw the
sun's image on to the surface of a boiler placed
at the focus.   The boiler, consisting of a coil of
blackened copper tubing, was 13 ft. 6 ins. in length,
with a capacity of 16 cubic feet, and operated under
a steam pressure of 150 lbs. per sq. in.   The reflector,
33 ft. 6 ins. in diameter, was rotated by clockwork
so as to keep its axis always directed towards the
sun.   This plant supplied a steam engine which
pumped water for irrigation purposes, and developed
from 6 to 8 horse-power.   The cost of the whole
installation was about £1000, or roughly twelve times
that of an ordinary steam plant to develop the same
power.

    This type of plant has many obvious disadvantages.
Its first cost per horse-power is excessively high even
for a small plant, and would be proportionately higher
for a larger plant owing to the strength of construction
which would be necessary to enable such an unwieldy
structure to withstand wind forces.   Furthermore,
owing to the smallness of the units and the somewhat
complicated nature of the mechanism, the cost of
supervision and of maintenance of a large installation
would of necessity be high.   After spending many
years and some £20,000 on his experiments, Ericsson

in fact came to the conclusion that for these reasons the system was impracticable on any commercial scale.

In order to avoid some of these disadvantages the direct-heating type of solar power plant has been favoured by some recent experimenters. In the earlier installations of this type no mirror is used, but the sun's rays are allowed to fall directly on the heat-absorber. This usually consists of a series of shallow trays or vessels, whose sides and bottom are insulated by some non-conductor of heat. They are covered by two closely-fitting sheets of glass with a shallow air-space between. Under normal conditions the sun's radiant heat falling on to the earth's surface is radiated or conducted away into the atmosphere as fast as it is received. This double layer of glass, however, largely prevents outward radiation and conduction, with the result that, just as in a green-house, the temperature inside becomes greater than that of the surrounding atmosphere. Water admitted at one end of the absorber flows through in a thin layer, absorbs heat, and is then pumped into an insulated storage tank, from which it is allowed to flow as required through the tubes of a tubular boiler containing ether, ammonia, or, more commonly, sulphur dioxide. This fluid is vaporized by the heat and is utilized to work a steam engine, after which it is condensed and again pumped into the boiler while the cooled water is readmitted to the cool end of the heat-absorber.

Such an arrangement is shown diagrammatically in Fig. 5.   Here H is the heat-absorber, S the storage tank, B the boiler, P_1 the pump for maintaining the

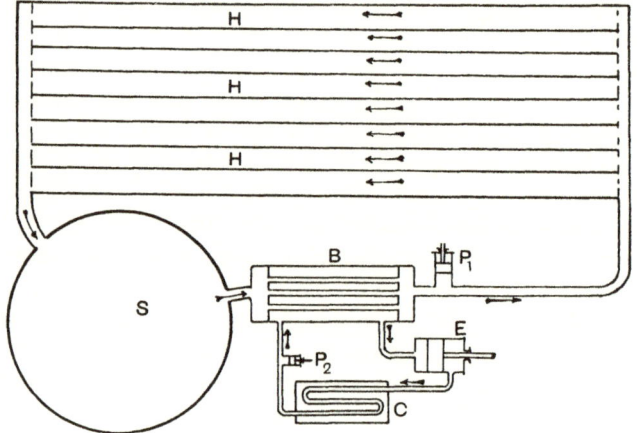

Fig. 5.   Diagrammatic sketch of Solar-Heat Plant with Absorber, H, and Hot Water Storage Tank, S.

circulation of water through the tank, E the engine with its condenser C and pump P_2 for returning the working fluid to the boiler.

In a plant of this type installed near Needles (Cal.) [10] in 1904, two heater sections were used, with a total surface of about 1000 sq. feet. The average temperatures attained under the glass of the heater during the month of June ranged from 120° F. at 7 a.m. to a maximum of 250° F. at 1 p.m., afterwards diminishing to 120° at 5 p.m., while the average temperature of the hot water entering the storage vessel was about 180° F. Sulphur dioxide was adopted as the working fluid in the engine, the working pressure being about 200 lbs. per sq. inch. As the result of tests, it was concluded that for a plant to be capable of continuous running, night and day, a heating surface of 400 sq. feet per horse-power would be necessary. The estimated cost of such a plant, including a storage vessel for hot water capable of storing 100 hours' supply, with the necessary vaporizer, engine, and condenser, is approximately £33 per horse-power, or about three times that of the average steam plant.

In a larger and somewhat different type of plant [11] tested at Tacony (Phil.) in August 1911, the absorber (Fig. 6 [12]) consists of a series of units each containing a flat metal vessel three feet square enclosed in a wooden box covered with two layers of glass separated by a one-inch air space. These boxes are insulated by a two-inch layer of granulated cork and are mounted on stands about 30 inches high, which permit them to be inclined perpendicular to the sun's rays at the meridian. Plane mirrors of cheap

Fig. 6. Heat-Absorber for Solar-Heat Plant.

construction, six feet apart at their upper edges, are mounted on two sides of the boxes in order that a larger proportion of the rays may be intercepted and reflected on to the surface of the absorber. Each water vessel is connected at one end to a feed pipe and at the other end to a steam pipe which conveys the low-pressure steam generated to a condensing reciprocating steam engine. The water from the condenser is pumped back into the absorber, so that the only loss of working fluid is that due to accidental leakage. Tests showed that in the neighbourhood of Philadelphia in August this absorber with a total area, including mirrors, of 10,300 sq. feet was capable of producing 4800 lbs. of steam per day of eight hours. In view of the low pressure of this steam and of the well-known steam-eating propensities of the small reciprocating engine, it is highly improbable that this would develop on the average more than 20 H.P. over the eight hours, or at the rate of 1 H.P. for each 500 sq. feet of surface.

In a more recent plant by the same inventor, recently installed in Egypt at Meadi, near Cairo [13], the design has reverted to the parabolic reflector throwing the sun's rays on to a closed metallic heater at its focus. In this plant five reflectors are used, each 204 feet long, and 10 feet wide at their upper extremity. These are built up of silvered glass mirrors mounted in a steel cradle of parabolic section.

The cradle is carried on a series of semi-circular racks which gear with pinions by which the reflector may be rotated so as to follow the sun. The heater, carried at the focus of the reflector, consists of a closed box of zinc plate, the same length as the reflector, 14 ins. high and $\frac{3}{8}$ in. wide. This is painted black over its outer surface. Water is admitted at one end while the other end is enlarged into a 4-inch steam pipe which conveys the steam generated to a low-pressure steam engine with surface condenser.

It is estimated that this plant is capable, in its present surroundings, of developing an average of 100 H.P. or roughly 1 H.P. for each 100 sq. feet of heating surface.

Taking the cost of a modern solar-heat plant at £33 per horse-power, and assuming the cost of attendance, maintenance, and stores to be the same as in a steam plant, the total cost of power would be roughly the same as that obtained from a steam plant consuming 3 lbs. of coal per H.P. hour, with coal at 12s. 6d. per ton. Probably this on the whole favours the solar-heat plant, inasmuch as its attendance and maintenance charges would be greater than in the average steam plant, while, unless erected on otherwise waste ground, the cost of the comparatively huge ground area necessary for the absorber would prove a severe handicap.

In tropical countries, where fuel is abnormally

expensive, the solar energy plant has obvious advantages, always provided its maintenance charges do not prove excessive, and future improvements and developments will undoubtedly do much to reduce these charges and to increase the reliability of such an installation. It is true that those parts of the earth's surface where a fairly steady supply of radiant heat can be relied upon are, on account of their excessive temperature and their lack of rain, quite unsuited to the maintenance of large centres of industry, while in temperate climates, or in tropical climates where the rainy season may cause a total stoppage of the power supply for some weeks on end, the method is obviously impossible.

Still with improvements in the method of transmission of high tension electricity over long distances the idea may be brought within the region of practical politics, and the amount of energy which might, theoretically, be rendered available in this way is enormous.

As indicated in Fig. 7, there is a rainless belt extending from the north-west coast of Africa to China, 8000 miles in length and some 800 miles wide, including the deserts of Northern Africa, Upper Egypt, part of Syria and of Arabia, the greater part of Persia and the western part of China. The southern portions of Africa, and in the western hemisphere a large portion of the United States and the western coast and central portions of South America are peculiarly well adapted for the utilization of radiant energy.

Making an allowance of 100 square feet per horse-power, calculations show that if only one per cent. of these areas were utilized roughly 20,000 million horse-power might be developed for eight or nine hours each day. Even if the only areas considered be those lying within five miles of tide water and thus readily accessible by water communication, the utilization of one per cent. of such areas would give roughly 1500 million horse-power, or enough, under present conditions, to do the whole mechanical work of the globe several times over.

Two-thirds of the Australian continent, with its area of 3,000,000 square miles, also suffers from continuous, intense radiant heat, and here alone, if one per cent. of a strip of its coast-line five miles wide were to be utilized, some 120 million horse-power, or 14,000 horse-power per mile of coast-line might be developed.

So long as such stores of energy are available, there is little reason to anticipate any insuperable difficulty in maintaining the industrial activity of the world as a whole. To our own and all northern communities, however, they can offer but little assistance even with all the help electric transmission can give, and their utilization on any large scale would

Fig. 7. Regions subject to intense Solar-Heat and with very slight Annual Rainfall.

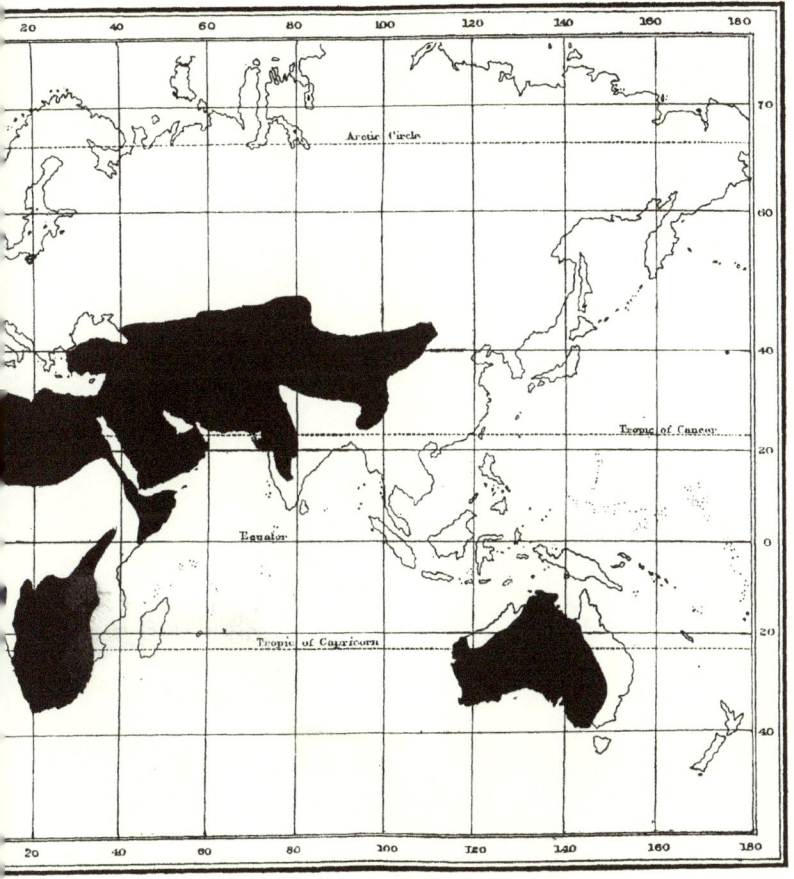

involve an immense change in the relative industrial importance of the countries of the world.

CHAPTER V

THE UTILIZATION OF VEGETATION FOR POWER PURPOSES

THE leaves of every form of vegetation are built up of cells containing a green colouring matter called chlorophyll. Under the action of sunlight this absorbs carbon dioxide from the atmosphere, and, in conjunction with water absorbed from the soil, builds up the carbon into the organic constituents of the plant. One of these products which enters largely into the composition of the fibrous structure is cellulose, a carbohydrate containing some 44 per cent. of carbon, 6 per cent. of hydrogen, and 50 per cent. of oxygen.

In many plants, such as the potato, and notably the cereals rice, maize and wheat, the carbohydrate appears mainly in the form of starch. In others, such as beet-root, a large percentage of sugar is formed. Either of these substances can be readily converted into alcohol. Recent experiments, indeed, show that alcohol can be produced by fermentation processes from such a vegetable material as wood sawdust to the extent of some 18 per cent. of its weight; and it is safe to say that when the necessity arises it will be found possible to manufacture alcohol from practically any form of vegetable matter.

Inasmuch as the chemical changes which occur in growing vegetation are indirectly due to the radiant heat and light of the sun, the use of such vegetation as a fuel affords another, although indirect, means of utilizing this radiant energy. Helmholtz has calculated that of the energy poured by the sun's rays on to any area of growing vegetation, about one fifteen-hundredth is utilized in producing photo-chemical reactions. Although this is an extremely small fraction of the whole, it is yet equivalent to some 200,000 B.TH.U. per minute per square mile of area exposed to the perpendicular rays of the sun, an amount of heat which, if it could be entirely converted into mechanical energy, would enable some 4600 H.P. to be developed per square mile for about eight hours per day.

It is estimated that the world's annual growth of vegetation amounts to some 32,000 million tons, equivalent as a fuel to about 14,000 million tons of coal, or about eleven times the present output.

This vegetation may be utilized as a fuel in either of two ways. That portion of it which takes the form of timber may be used as a boiler fuel, or in a gas producer to produce power gas. In the latter case it may either be used in its natural form, or may first

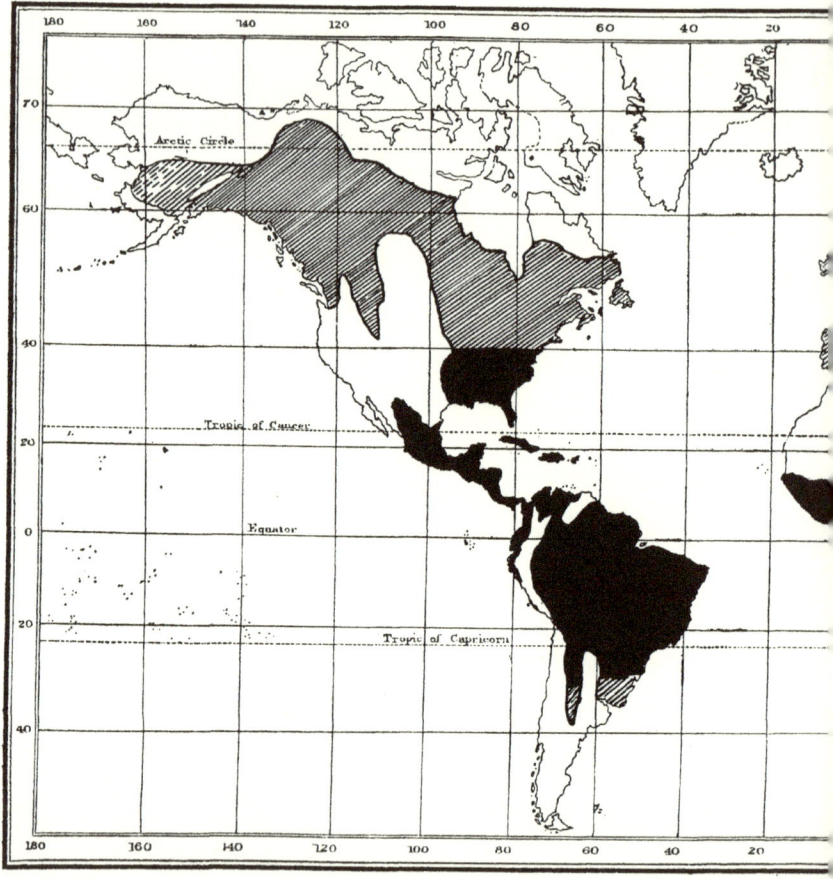

Fig. 8. Regions suited for the maintenance of Vegetable and Plant Life. Luxuriant vegetation shown in black

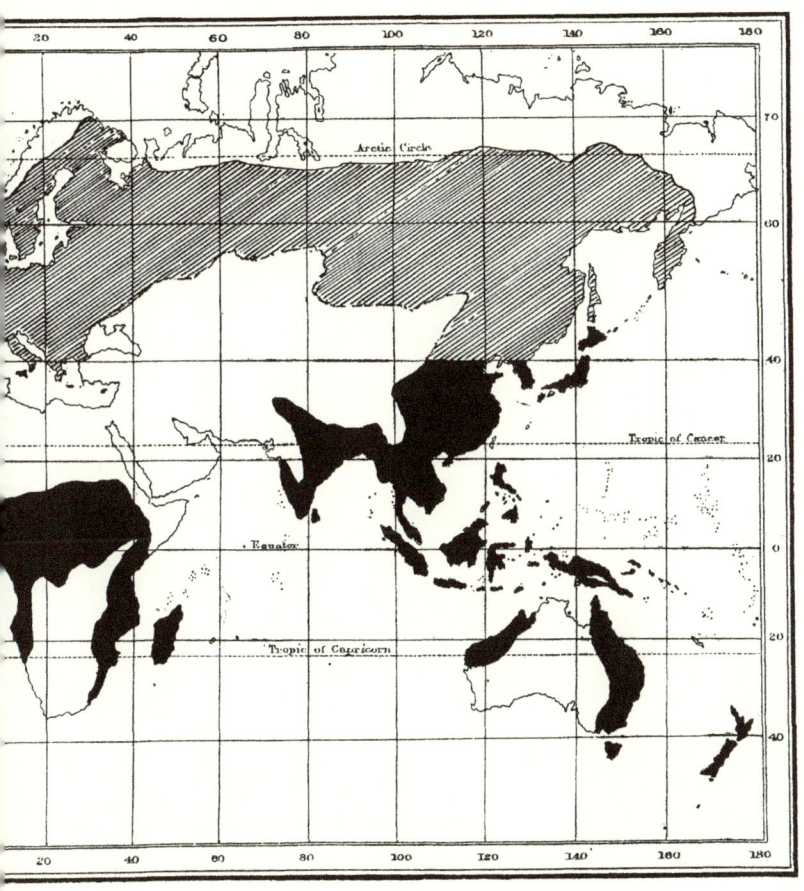

be carbonized into charcoal, which may afterwards
be used in a gas producer, while the tar given off in
the process may be used as an oil-engine fuel. Other
and smaller forms of vegetation may, if necessary, be
air-dried and used in the same way, but in view of
the large percentage of moisture which they commonly
contain, these can generally be more profitably utilized
for the manufacture of alcohol. By gasifying the fuel,
or the waste fibre left over from the manufacture of
alcohol, in a producer with ammonia recovery, a very
complete cycle is maintained with only a comparatively
slight waste of energy. The ammonia is returned to
the soil as a fertilizer along with the mineral substances
in the ash, while the carbon dioxide from the combustion of the fuel gas is returned to the atmosphere
to be again utilized by growing vegetation.

Timber. The possibility of utilizing timber as a
partial substitute for coal as fuel is worthy of close
consideration. Mr J. C. Hawkshaw in his presidential
address to the Institution of Civil Engineers in 1902,
estimated that the wood-fuel produce of an acre of
land in Europe is equivalent to at least one ton of
coal a year, and provided that each tree as felled is
replaced by a seedling, this rate of supply is continuous and permanent. On this assumption two
million square miles of forests would be required
to give a continuous fuel-supply equivalent to the
present coal output of the world. As the land area

of the globe is in the neighbourhood of 50 million square miles, such forests would only cover some four per cent. of its area. Even allowing for the fact that probably some sixty per cent. of the total area is unsuitable for timber growing or can be more profitably utilized in other ways, the maintenance of such an area does not appear to present any insuperable difficulties. On the other hand, at the present time the world's output of timber is barely sufficient to satisfy its demands for purely constructional and industrial purposes, and in view of the magnitude of the problems involved in the maintenance and handling of timber in such bulk it is extremely improbable that the utilization of this material as fuel, on any very large scale, will ever become a practical proposition. The cost of such a fuel would of necessity be high compared with the present price of coal or oil. On a comparatively small scale, however, the extension by afforestation, and the utilization of the timber supplies in suitable localities are to be anticipated.

In our own country the possibilities in this respect are not great. At present some three million acres, or about five per cent. of the total area of the British Isles, is under timber, and this, as a fuel, would give annually the equivalent of only about one per cent. of our coal output. Even although this acreage might readily be doubled or even trebled, the resultant output would be insignificantly small.

The value of wood as a fuel depends largely on the amount of moisture which it contains. When newly felled this ranges from 30 to 50 per cent., and after 12 months' air-drying is reduced to 20 or 25 per cent. When well air-dried, the heat of combustion of all woods is approximately the same, one pound of coal being equivalent as a fuel to about 2·25 lbs. of wood.

When made into charcoal by distillation in a retort some 30 per cent. of the weight treated is returned in this form, and as about 12 per cent. of the weight is burned in the furnace during the process, the proportion actually converted into charcoal is only about 27 per cent. of the whole. As a fuel, charcoal has about double the heating-value of air-dried wood, so that only about one-half of the heat-value of the original material is available in the charcoal produced.

Alcohol. In many respects the prospect of utilizing the resources of the vegetable kingdom for the production of alcohol is more promising. Although the cereals contain a much greater proportion by weight of starch (wheat about 60 per cent., rice 83 per cent.) than do potatoes (about 20 per cent.), the weight of starch obtainable per acre is much larger with the latter crop. Thus, whereas an acre of grain will yield some 900 lbs. of starch, the same area under potatoes will yield about 2500 lbs., and this, when used for the manufacture of absolute alcohol, produces about 1400 lbs. of the spirit.

The heat-value of this is 12,600 B.TH.U. per lb., or roughly twice that of air-dried wood, so that the heat-value of the alcohol produced per acre per annum is equivalent to that of 1·25 tons of wood, or about one-half that of the timber grown on the same ground.

When it comes to converting this heat into mechanical energy, however, the two processes become more nearly equal. Used as a fuel in an internal combustion engine, the alcohol is capable of converting some 35 per cent. of its heat into mechanical energy. The wood, if used as a boiler fuel, only utilizes some 12 per cent. of its heat, while if used in a producer in connection with a gas engine, this may be increased to about 22 per cent. In the latter case, a given acreage devoted to the growth of timber will give about 25 per cent. more power, while in the former case it will give about 30 per cent. less power than if utilized for the production of alcohol.

As a fuel for internal combustion engines, alcohol is capable of giving excellent results, and in point of safety, thermal efficiency, and flexibility, has some advantages over petrol or kerosene, particularly for comparatively small powers. It compares unfavourably with petrol in corrosive effect ; in the difficulty of starting up an engine cold; and in the fact that external heat is required to vaporize it.

As regards corrosive effect, this is probably due to impurities in the alcohol as prepared for commercial use. Experience, however, shows that no appreciable corrosion takes place even with such alcohols, except on those parts of the motor which are so cold as to produce condensation, and by keeping the jacket water at a sufficiently high temperature, all condensation and corrosion in the engine itself is readily prevented.

The difficulty experienced in starting with a cold engine may readily be overcome by the addition of a little benzol to the alcohol, or by the provision of means for preheating the carburettor before starting. It is true that alcohol requires external heat to vaporize it, and to this extent its possible thermal efficiency as a fuel suffers. When used in an internal combustion engine this drawback is, however, more apparent than real, inasmuch as a large proportion of the total heat of combustion must pass away in the exhaust gases, and this, which would be otherwise wasted, is readily available for the purpose of vaporization.

When used for power purposes, alcohol at 3*d*. per gallon is equivalent to coal at about 35*s*. per ton, or to petroleum at about £5 per ton. Evidence shows that when produced from sawdust or from peat a figure as low as this may be attained. Using potatoes as a base, the cost, including denaturing, is of the order of 1*s*. per gallon, while if made from beet-root it is a little less than this. There can be little doubt, however, that any great demand for alcohol as a fuel would bring other and cheaper vegetable materials forward as a base for its manufacture, and would result in a cheapening in the costs of the manufacturing processes, in which case any appreciable increase in the present cost of coal or of oil would enable it to compete on level terms as a commercial proposition for even large powers.

Vegetable Oils. In addition to producing woody fibre and starch which may be directly or indirectly used as fuel, many plants produce fruits or seeds from which vegetable oil may be obtained by pressure. Among such products are castor-oil, palm-oil, and earthnut or peanut oil. Recent experiments [14] on peanut oil show that this is well adapted for use in the Diesel engine, and that its consumption only amounts to about ·53 lb. per B.H.P. hour—or practically the same as with petroleum. Similar successful experiments have also been made with castor-oil as a fuel. As the plants for producing these oils can be readily grown in considerable quantities in most tropical climates, this affords yet another promising method of utilizing the energy of plant-life for power purposes.

Wonderful as is the process of photo-chemical synthesis by which the cell matter of the living plant with the help of solar energy builds up carbohydrates of considerable complexity from carbon dioxide and water, recent experiments by Prof. Ravenna and Giacomo Cimician at Bologna, show that this process may be modified artificially to a certain extent, and that a plant may even under certain conditions be made to synthetise a substance which it does not usually produce. For example, maize was forced to yield a glucoside (from which sugar may be produced), while the production of nicotine in the tobacco plant was largely increased or reduced as required. As yet the work is not sufficiently developed to enable an estimate of its possibilities to be formed, but it is by no means inconceivable that by modifications in the synthesis of such a plant as the potato or beet-root, the output of alcohol per acre might be very considerably increased.

CHAPTER VI

THE INTERNAL HEAT OF THE EARTH

It is a matter of common knowledge that the temperature of the earth increases towards its interior, and the suggestion of utilizing this internal

Fig. 9. "Old Faithful" Geyser, Yellowstone Park. Height, 165 ft. Eruptions lasting for 7 minutes at intervals of about 70 minutes.

heat has long been a favourite one. At first sight this source of energy appears most promising. At certain points of the earth's surface, notably in the Yellowstone Park, in Iceland, and in the northern districts of New Zealand, abundance of water at a temperature of about 200° F. is brought to the surface by hot springs and geysers in close proximity to supplies of water at about 60° F. In such cases there is no reason why this difference of temperature should not be utilized in the working of a steam or vapour plant. It is true that the higher temperature is somewhat low to enable steam to be economically used, but it is sufficiently high for a motor using either ether vapour, sulphurous acid, or carbonic acid gas as its working fluid. In such a plant the evaporator would receive its heat from the hot water supply while the condenser would be cooled by the adjacent cold supply. Points so favoured are, however, very rare, and the horse-power which they might render available would not amount in the aggregate to more than a few thousands.

In a scheme for utilizing the earth's internal heat at any other point of its surface, it would be necessary to sink two or more bore holes to a depth sufficiently great to enable the required high temperature to be attained ; to connect these at their lower extremities; and to circulate a continuous stream of water down the one and up the other bore hole. The water, heated during its passage, would then be utilized as already described.

As thus baldly outlined, the method appears simple and perfectly feasible, but a closer examination shows that in general the difficulties are so great that this source of energy must be definitely abandoned as a serious factor in the future power problem.

Much information as to the rate of increase of temperature with depth below the earth's surface, has been rendered available by deep borings which have been made during recent years. Under normal conditions it is found that the temperature at depths less than 100 feet suffers annual seasonal fluctuations, and that below this depth the increase of temperature with depth is of the order of 1° F. for each 60 or 70 feet in depth.

This number, however, varies within fairly wide limits at different parts of the earth's surface. In favourable localities, in the neighbourhood of volcanoes or where the earth's crust is known to be thin, the rate of increase is of course much greater. In the case of a railway tunnel driven through the base of Vesuvius the normal temperature of the working atmosphere was between 160° and 170° F., while in a hole made in the newly exposed rock a maximum temperature of 200° F. was registered. Unfortunately such localities, on account of the possibilities of earthquakes and of volcanic eruptions, are not desirable places for the establishment of large centres of population or of industry.

At Wheeling, West Virginia, the temperature gradient as shown by observations in a bore hole 4500 feet deep is about 1° F. per 80 feet over the upper, and about 1° F. per 60 feet over its lower portion. The temperature at the bottom is about 106° F.

In a second well near Pittsburg, the depth of which is 5580 feet, practically the same temperature gradient is attained, while in the old Comstock Mine at Virginia City, Nevada, the gradient is so rapid that temperatures of 170° F. are reached in the workings.

In a bore hole at Cynchous (Silesia)[16], 7347 feet deep, the average temperature gradient is 1° F. for each 55 feet. In this well, moreover, the gradient is much more rapid (1° per 31 feet) over the middle third than over the lower third, where it is only 1° per 91 feet. The probability is that this rapid gradient near mid-depth is due to oxidation in the surrounding material, or possibly to the presence of some radioactive deposit, and that in a comparatively short time the gradient will become more sensibly uniform over the whole depth. The temperature at the bottom of this bore hole is 182° F.

On the other hand, rock borings 1000 feet deep on the Witwatersrand [17] only show a rise in temperature of 1° for each 208 feet of depth, the temperature at 1000 feet not exceeding 69° F.

On the assumption of a mean annual temperature of 65° F. at a depth of 100 feet and of a gradient of 1° F. per 56 feet afterwards, the following table shows the temperatures attained at different depths.

Depth (miles)	½	1	2	4	6	8
Temp. °F.	103	143	223	383	543	703

Thus in order to attain a temperature equal to that of water boiling at atmospheric pressure, a depth of practically two miles would be necessary.

Even assuming these temperatures at the bottom of the bore holes to be initially attained, the temperature of any steady stream of water delivered from these holes would of necessity be very considerably lower. Directly the flow of water commenced the temperature of the surrounding rocks would be lowered and finally, when a steady state was attained, only as much heat would be given to the water as could be transmitted by conduction through that portion of the earth's strata surrounding the bore holes and the lower reservoir.

Thus it would be necessary to drive to a depth having an initial temperature of four or five hundred degrees in order to obtain a working temperature of two hundred degrees. The final temperature would depend entirely on the volume of water handled ; on the size of the reservoir at the bottom of the bore holes ; and on the heat conductivity of the rock.

In order to obtain as large a reservoir as possible the suggestion has been made of shattering the rock between the bore holes by charges of nitro-glycerine. Unfortunately, however, in order to get a cavity of any appreciable size, the displaced material must of necessity be removed, and this necessitates one of the holes being of sufficiently large dimensions to serve as a working shaft for this excavation. In his presidential address to the British Association in 1904, the Hon. C. A. Parsons gave the following as the estimated costs and times necessary for the construction of such shafts.

Depth	Cost	Time in years
2 miles	£500,000	10
4 ,,	£1,100,000	25
6 ,,	£1,800,000	40

It is true that modern engineering methods would enable both cost and times to be substantially reduced from these estimates, but still they would remain comparatively enormous.

A little calculation enables one to estimate what would be the probable output of such a system. Assuming, under steady working, that in the rock surrounding the cavity the temperature gradient was as high as 1° F. per 10 feet, we should get a supply of heat by conduction, of about 0·12 B.TH.U. per square foot of surface per hour. If then the cavity had a diameter of say 300 yards, with an internal area of 283,000 square yards, the available supply of heat would only be sufficient to develop, with an ordinary engine, about 15 horse-power continuously, certainly an absurdly inadequate return for the energy expended in sinking the bore holes alone. Furthermore no account has been taken of the difficulties which would be experienced in ensuring an effective circulation of the water over the surface of the heating chamber. In view of these considerations it would appear that except in the few localities before mentioned the direct utilization of this energy on any commercial scale is quite impracticable, and that even in these localities the comparatively shallow bore hole with its medium temperatures offers much greater commercial possibilities, than does the deep bore hole with its higher temperatures.

It should, however, be realised that in virtue of its conduction to the earth's surface, this energy is being continually utilized in the production of food stuffs, and of timber, and in the evaporation of surface moisture, and so must, albeit indirectly, always play a large part in the energy scheme of the globe [18].

CHAPTER VII

WATER POWER

THE mean annual rainfall of the world is about
36 inches per annum, giving an average precipitation
of some 80 million cubic feet per square mile per
annum on the land of both hemispheres. Taking the
mean available height of fall as 50 feet, and assuming
it possible to collect and store fifty per cent. of the
total rainfall in reservoirs this would give roughly
120,000 million foot-pounds of available energy per
square mile of gathering ground per annum, and this,
if expended in 3000 working hours in hydraulic
motors having an efficiency of 80 per cent., would
give a total of 16 H.P. per square mile in such
districts as might be able to utilize such a fall.

It is true that in many localities the head which
may be made available by suitably placed reservoirs
is much greater than 25 feet, and in specially favourable situations may amount to many hundreds of feet.
Such, however, are the exception, and those regions
which in virtue of their steep gradients associated
with heavy precipitation are rich in water powers,
are not usually otherwise suited to the support of a
large population. Still in the United States alone
such sources of energy, variously estimated to aggregate between 35 and 55 million H.P., are known to be

Fig. 10. Niagara Falls.

capable of commercial development. Of these some
5·3 million H.P. were actually utilized in 1908.

This total is very largely increased by water
powers known to exist in other parts of the world.
In this respect South America, Canada, portions of
British India, of South Africa, and of Europe are
favourably circumstanced, as indicated in Fig. 11,
and it is safe to say that those water powers which
may be successfully utilized when the necessity arises,
amount to well over 200 million H.P. or more than
enough to provide mechanical energy for the whole
present work of the world.

In Europe the bulk of the powers are small, yielding from a few hundred to a few thousand H.P. each,
but in spite of their moderate size the wideness of
their distribution makes them specially valuable.
Little has as yet been done to develop the hydraulic
resources of the United Kingdom, and indeed they
are comparatively very small. Situated mainly among
the Highlands of Scotland and Wales it is doubtful
whether they aggregate as much as one million H.P.

Rapidly as the large water powers of the world
are being developed at the present time, comparatively
few of them supersede the use of fuel. They go rather
to building up new industries and so far there is little
sign of hydraulic developments checking the demand
for fuel.

Although the question of the utilization of natural

water powers has always been of great economic
importance, the introduction and perfection of electrical methods of high-tension transmission of energy
for great distances with little loss, have made it
possible, of recent years, to take advantage of such
powers far remote from large centres of industry, and
have greatly enhanced their commercial value. Where
not handicapped by the necessity for costly constructional works, such installations can now be made to
accomplish even electrical heating at a cost not
greatly in excess of that of heating with coal as a
fuel, and the time is not far distant when every
cataract of size, whether near or remote from the
industrial haunts of man, will be harnessed and made
to do its share in the energy scheme of the globe.

The available energy of any elevated store of
water is measured by the product of the head H and
of the weight of water which may be utilized per unit
time. If the fall be measured in feet and if the
quantity is Q cubic feet, or $62\cdot4Q$ lbs. per second, the
energy available per second is $62\cdot4QH$ foot-lbs., equivalent, in a perfect hydraulic turbine, to $\dfrac{62\cdot4QH}{550}$ H.P.
Actually the efficiency of a turbine does not exceed
some 80 to 85 per cent., while if this be utilized to
drive an electric generator which in turn supplies
current to drive an electric motor, the combined
efficiency of generator and motor will not usually

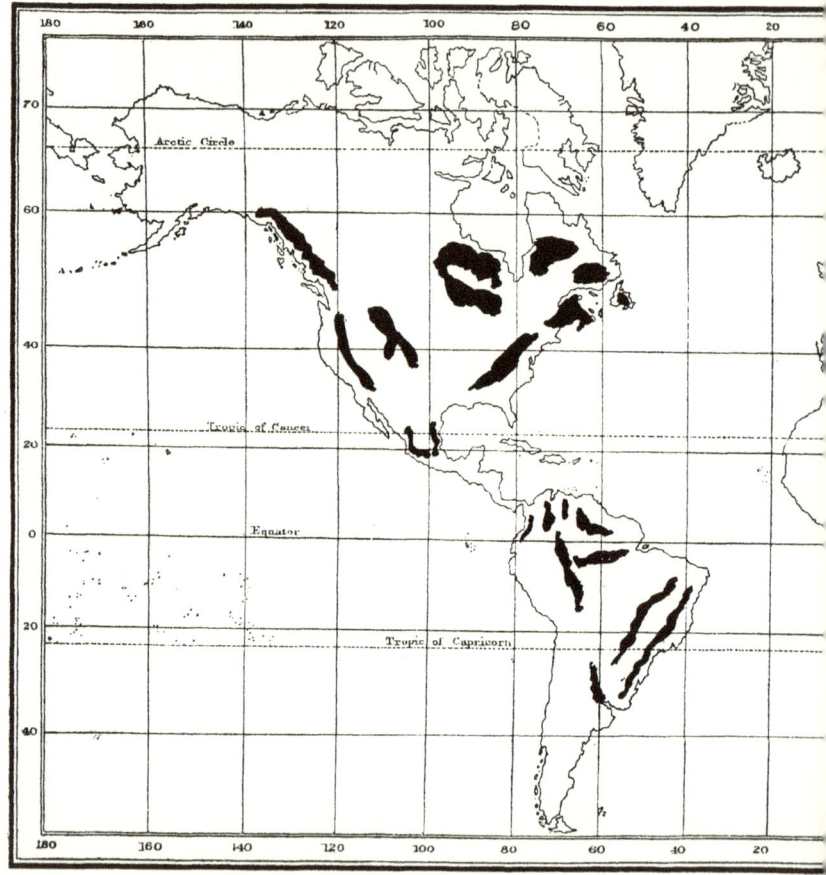

Fig. 11. The Principal Water Powers of the World.

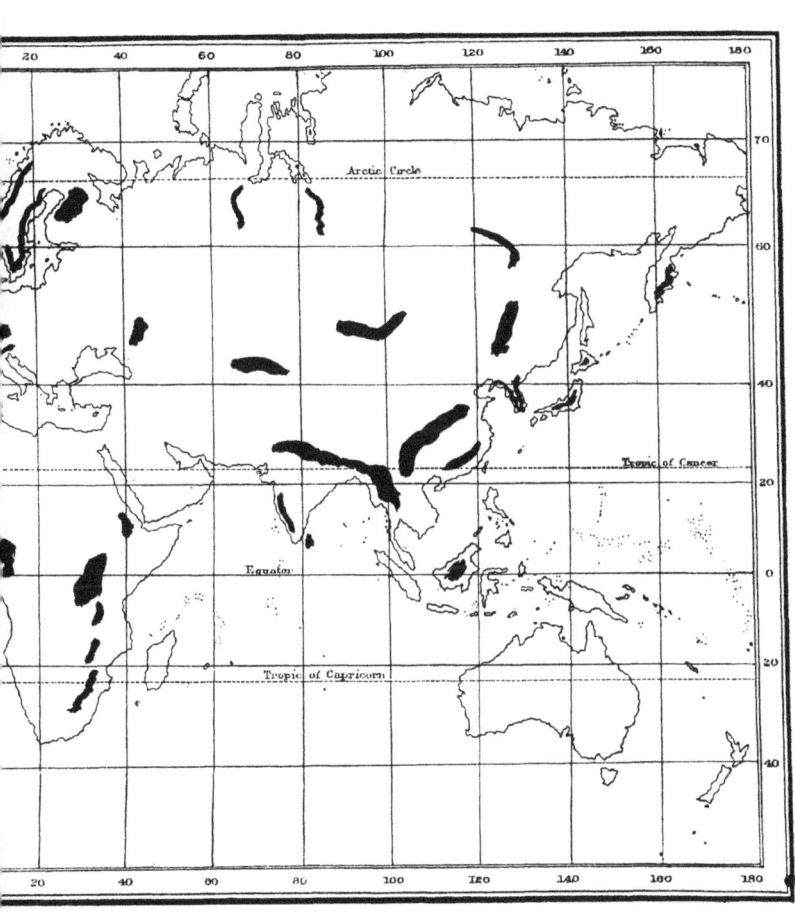

exceed about 82 per cent.   Thus when losses due to
transmission between generator and motor are taken
into account, the energy actually delivered by the
latter in the form of useful work will not exceed some
60 per cent. of the energy of the water supply.

As the usefulness of a water supply depends in
most cases on the possibility of maintaining its
uniformity over long periods of time, the maximum
useful power is strictly limited by the minimum
power which, with the aid of any suggested storage
system, will be available towards the end of the
longest probable period of drought.

This can only be predetermined in any given case
from a knowledge of the discharge from the suggested
gathering ground or catchment area, extending over
a long series of years.   Where such a record is not
available a close investigation of the rainfall records
of the area, or of adjacent areas having the same
general physical situation and characteristics, will
enable the minimum probable rainfall under the
worst combination of circumstances to be calculated.
When this is known simultaneous measurements of
the rainfall over the catchment area and of the
discharge of the various streams draining it, enable
the ratio of rainfall to run-off to be determined, and
the minimum probable monthly or annual run-off can
then be calculated.

Of the rain which falls in any watershed, part is

evaporated by the direct heat of the sun; part absorbed by the soil and the leaves and roots of vegetation ; while part finds its way into some spring or stream, and is known as "run-off." The proportions of the rainfall which are utilized in these various ways vary greatly with the physical characteristics of the ground and locality.

The effect of evaporation is much more marked in flat than in hilly districts, and in hot and dry climates than in those more temperate. In a very hot and dry climate evaporation from an open sheet of water amounts to as much as 8 feet per annum, while in temperate climates in hilly districts it may be as little as 10 inches per annum. Generally upon permeable soils or upon steep and impervious land this loss is small.

A given amount of precipitation concentrated in a few heavy showers will give a greater run-off than the same amount falling in a continuous but mild downpour. Also when the soil is baked by a drought, a larger portion will run off than when it is slightly moist. Rain falling at night gives a larger run-off than that falling during the daytime owing to reduced evaporation, while rain falling on frozen ground appears almost wholly as run-off. The discharge or run-off from a catchment area thus varies within wide limits daily, monthly, and yearly, and only bears an indirect relationship to the rainfall.

Investigations on the Sudbury (Boston) watershed [19] during the years 1875 to 1890 show the
following run-off, expressed as a percentage of the
rainfall.

Period	Max.	Min.	Mean
January	88·8	7·6	49·1
February	206·9	42·9	78·2
March ...	261·7	74·0	109·6
April ...	188·3	48·5	109·1
May ...	260·2	39·9	62·3
June ...	54·7	14·0	29·3
July ...	20·9	3·6	8·9
August ...	61·2	4·1	13·0
September	30·9	6·1	14·2
October	71·4	4·8	23·1
November	174·7	11·3	39·5
December	127·3	9·6	52·5
Mean	62·2	31·9	49·5

This watershed has an area of 75 sq. miles, and is
hilly, with some large swamps within its borders.

In this country, gaugings of the river Lee during
the years 1851, 1852, 1856, and 1858 gave the following results.

Year	Rainfall (inches)	Run-off (inches)	Percentage run-off of rainfall
1851	22·62	6·00	26·5
1852	39·71	9·13	22·7
1856	23·91	5·57	23·3
1858	20·86	4·39	21·0

The rainfall of a district varies greatly with its situation, physical configuration, and altitude, and with the direction of the prevailing winds. Where these are charged with moisture through crossing a large extent of open water, the rainfall of the first high ground encountered by them will be heavy.

The west coast of Scotland, the north-west coast of England, the New England States, and Northern California afford examples of this. The annual rainfall in the former districts often exceeds 80 inches while at Seathwaite in Cumberland it averages about 150 inches. In the New England States it averages from 50 to 60 inches and in Northern California from 40 to 50 inches.

On the west coast of India between latitudes 10° and 20° north it often attains 250 inches, and in isolated districts double that amount. On the other hand, the rainfall of a district is small if the prevailing winds traverse a large expanse of land before reaching it, or if they come from a district of high to one of lower elevation. Thus the flat districts east of the Pennine Chain in England have a rainfall not much in excess of 20 inches annually, while the middle portions of the United States between longitudes 100° and 120° west have a rainfall varying from ·10 inch to 20 inches.

The method of utilizing a water supply depends largely on its magnitude and form. Where it takes

the form of a large river its flow may be diverted by
dams or weirs at each suitable fall in its level, and
the water may then be passed through a series of
turbines before being returned to the river. Under
such circumstances no method of storing the excess
rainfall in wet seasons for use in times of drought is
practicable, and the minimum discharge at such
times determines the maximum power which can be
developed continuously. This gives the cheapest
form of hydraulic power installation.

Where the source consists of a comparatively
small stream with a fairly constant flow throughout
the year, this may be utilized in the same way and
subject to the same restrictions as the large river.
But where, as is commonly the case, energy is only
required for the 8 hours or so comprising a working
day, the formation of a small reservoir capable of
impounding a 16-hours inflow will enable energy to
be utilized during the working day at a rate three
times as great as the mean rate of inflow.

Where the supply is very variable from month to
month, a still larger reservoir enables the wet weather
excess to be stored against the requirements of a
time of drought, but although desirable and indeed
essential in many cases, the necessary cost of land
and of constructional work in connection with such
reservoirs greatly increases the ultimate cost of the
power developed. In fact in many cases the cost of

such storage facilities and of the constructional works
necessary to bring the water from the reservoir to
the turbines ; to regulate its flow ; and to discharge
it to waste, amounts to 75 per cent. of the total cost
of the installation.

Hydraulic prime-movers.   The first hydraulic
prime-mover was a wooden paddle wheel, with its
paddles dipping into, and driven by, the current of a
stream, and as such a machine was at first only
required to do the work previously performed by
manual labour the power required was small and the
efficiency of little importance.   The construction of
these early wheels was of the most primitive type,
but was gradually improved.   Iron replaced wood ;
improvements in design led to increased efficiency ;
the demand for larger powers led to the utilization
of higher falls and to the consequent development of
a machine in which the water was admitted to the side
or top of the wheel, until a type of water-wheel was
evolved which, within its limitations, was as efficient
as the most modern of turbines.   Its chief disadvantages lay in its slow speed of rotation, the
impossibility of close speed regulation, and in the
large size of wheel required for even small powers.

For the purposes for which it was first required
these drawbacks were not serious, but the introduction
of more modern machinery involved the necessity for
a motor which, having a fairly high speed of rotation

in order to avoid excessive gearing losses, should be capable of close speed regulation and of utilizing higher falls and large volumes of water. For such purposes the water-wheel has been almost entirely superseded by one or other type of turbine.

The Pelton Wheel. Although the ordinary water-wheel is unsuitable for large powers, or for utilizing a head of more than twenty or thirty feet, one form of water-wheel—the Pelton Wheel—is well adapted for large powers and for the highest heads. In fact for heads of more than about 400 feet, for units developing less than about 1000 H.P., it will in general be found to be the most suitable form of motor, while for units up to 10,000 H.P. it is for many purposes to be preferred to its only serious rival, the inward radial-flow turbine.

In an installation of this type, water is led from the elevated storage reservoir through a pipe line terminating in one, or sometimes two, nozzles, which play on to a series of buckets fixed around the periphery of the wheel. In a modern wheel the buckets are of the general form indicated in Fig. 12. Each has a sharp-edged central ridge which catches the impinging jet and divides it without splash into two portions which are deflected backwards by the concave surfaces of the bucket. The peripheral speed of the wheel is about one-half the spouting velocity of the jet, so that the absolute velocity of the water leaving the buckets is very low, and, in a well designed wheel, when the water drops out of the buckets

Fig. 12. Disc and Buckets of Pelton Wheel.

into the tail race it has converted some 90 per cent.

of its initial energy into work. The supply of water
to a Pelton Wheel is usually adjusted by means of a
needle-regulator. This consists of a cylindrical needle
of tapering section fitted inside the nozzle, axially
with the jet. The water flows through the annulus
between the needle and the nozzle and forms a solid
cylindrical jet on leaving the needle. The latter is
capable of axial motion in the nozzle and may be so
adjusted as to fill the orifice either wholly or partially.
Its position, and therefore the size of jet, is regulated,
either by hand, or, more usually, by means of a
hydraulic relay cylinder which is operated through
the governing mechanism, which thus regulates the
supply of water to the wheel in accordance with the
demand for power.

Such an installation is eminently well adapted for
driving high-speed electrical machinery; is capable
of very accurate speed regulation; and has an overall
efficiency which for all powers from one-half to full
load is in the neighbourhood of 80 per cent.

The Pressure Turbine. Where it is desired to
develop a large power in a single unit with a comparatively low head of water, the volume to be
handled is of necessity large, and in a Pelton Wheel
this would only be possible with jets of large diameter
or with a number of jets playing on to a single wheel.
In practice it is found that if more than two jets are
fitted these interfere with each other to an extent

which seriously reduces the efficiency, while the diameter of the jets cannot, for various reasons, be increased to much above six inches. Where this would not enable the requisite power to be obtained, the Pelton Wheel is unsuitable and some form of so-called Pressure or Reaction turbine is used.

Such a turbine differs from a Pelton Wheel in two important respects. In the first place it is always so designed as to allow of water being admitted simultaneously at all points on its circumference, thus enabling a much greater volume of water to be handled and a much greater power to be developed with the same size of wheel. In the second place, in the Pelton Wheel the whole head of the supply water is converted into kinetic energy in the nozzle before it enters the wheel, and the water during its passage through the buckets is thus everywhere at the same (atmospheric) pressure. In the Pressure Turbine, on the other hand, the water leaving the guide vanes and entering the wheel is under pressure and thus supplies energy partly in the kinetic form and partly in the form of pressure energy. In its passage through the wheel this pressure is gradually converted into kinetic energy and the water is finally discharged at a pressure not sensibly different from that of the atmosphere.

In a pressure turbine the general direction of flow through the wheel may be either radial, or

parallel to the axis. If radial it may be either in-
wards or outwards. The first pressure turbines were
of the axial flow type. These were largely replaced
by machines of the outward radial flow type, which,
in turn, have given way to the turbine having inward
radial flow. All large modern machines are of the
latter type, modified in most cases so as to give an
axial, or partially axial discharge, and this type of
turbine—called after its inventor the "Francis"
turbine—marks what is at present the highest development of turbine design.

Such a machine consists essentially of a ring of
fixed guide passages, or vanes, which surround the
periphery of the wheel or runner, and guide the
supply water into a direction more or less tangential
to the latter. The runner itself carries a series of
vanes which are so designed as to receive, without
shock, the water leaving the guide vanes, and to
change its direction, finally discharging it into the
discharge pipe with zero tangential motion. In the
process its tangential momentum is destroyed, and it
is this change of tangential momentum which gives
rise to the turning moment on the wheel.

Figs. 13 and 14 show the construction of such
a machine, having a runner 5 ft. 6 ins. in diameter,
and designed to develop 10,000 H.P. at 300 revs. per
minute, under a head of 260 feet. In this turbine
the guide vanes are mounted around the inner

periphery of an annular spiral casing to which the

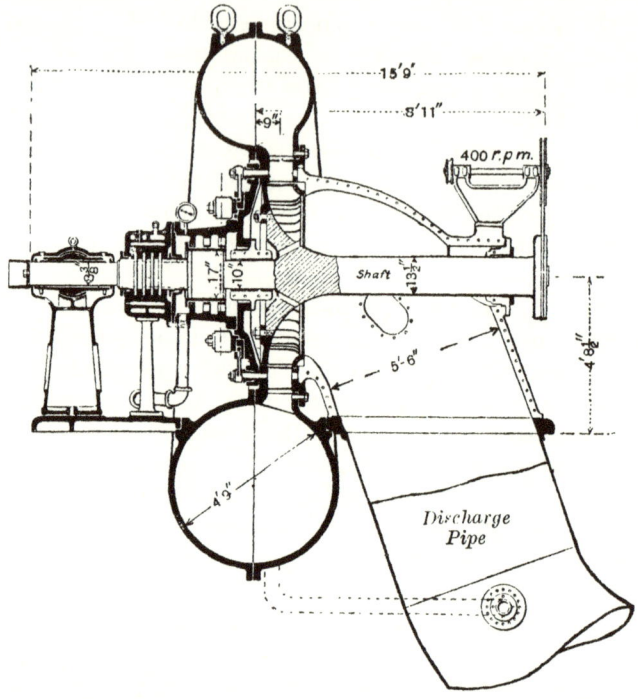

Fig. 13. Longitudinal section of Single Wheel Francis Turbine,
developing 10,000 H.P. at 300 revs. per minute, under 260 feet
head.   Diameter of runner, 66 inches.

pressure water is supplied from the supply pipe.   These

guide vanes are not fixed, but are pivoted so as to be capable of rotation about their own axes. Each vane is coupled, by means of a short lever mounted on its

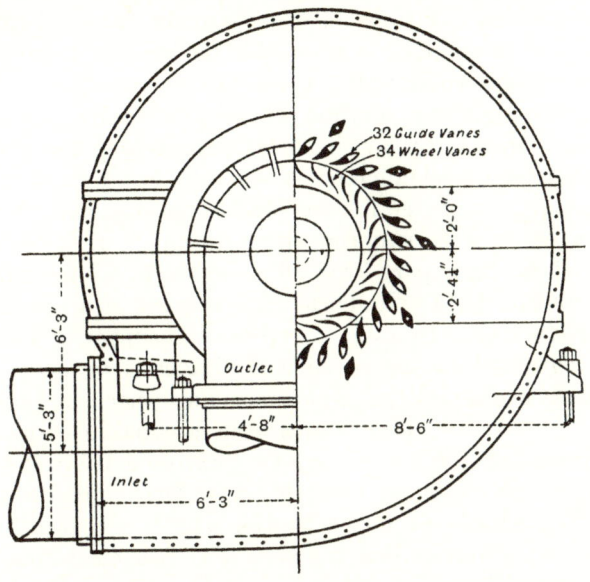

Fig. 14. Cross Section and End Elevation of 10,000 H. P. Francis Turbine.

pivot, to a movable ring which is outside the turbine casing and is concentric with the turbine shaft. This ring is rotated by means of a rack-and-pinion gearing

7—2

which is actuated by the governing mechanism. Rotation of the ring produces a rotation of each guide vane and thus increases or diminishes the effective area of waterway between adjacent guides. The turbine itself has only one bearing but is directly coupled to an electric generator having two bearings, making the whole unit a three-bearing machine.

Although the method of regulation by pivoted guide vanes is by far the most satisfactory and gives higher part-load efficiencies than any other method, a machine thus fitted is somewhat expensive, and where a machine is to run almost continuously at full-load, or where high part-load efficiency is not important, regulation by a cylindrical gate permits of a cheaper construction. In such a turbine the supply of water is regulated by means of a sliding ring or cylinder which works in the small annular space between the wheel and guide vanes and is capable of axial adjustment. When in its two extreme positions it respectively entirely cuts off, and offers no resistance to, the supply of water to the wheel. Its working position, intermediate between the two extremes, is regulated by the governor, and the system affords a very simple and effective means of regulation whose chief drawback is that it is very wasteful of energy at part-loads.

In spite of the many advantages of the Francis turbine, its high first cost prohibits its use in many

instances, and the demand for a cheap turbine suitable for medium falls has led to the development of
a type of machine known as the "mixed flow" or
"American" type turbine. In such a turbine the
ratio of depth to diameter of the runner is much
greater than in the normal Francis turbine. As in

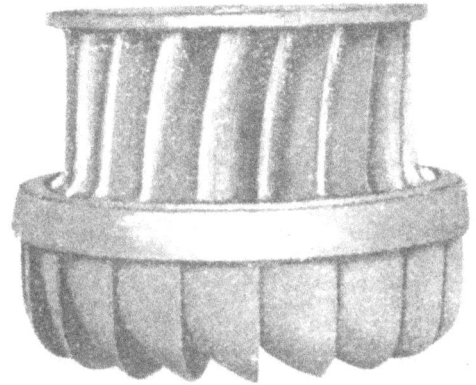

Fig. 15.   Runner for American Type Mixed Flow
Pressure Turbine.

the latter machine, water enters the wheel with
inward radial flow. After the inlet the buckets are
however curved both laterally and vertically. The
water in its passage through the wheel traces out a
path which is approximately a quadrant of a circle,

and is finally discharged partly in an axial and partly in an outward radial direction.

Fig. 15 shows such a wheel as fitted to a modern turbine of this type.

As thus constructed, an extremely large inlet and outlet area may be obtained with a runner of comparatively small diameter, and the volume handled and the power developed are large in comparison with the first cost. Although at full power the efficiency of such a machine is little less than that of the more expensive Francis turbine, its part-gate efficiency is usually considerably less.

In order to prevent flooding of a turbine installation by the banking up of the tail race water in times of flood it is usual to place the power house at a sufficient elevation above the tail race. If the water were discharged directly from the turbines at atmospheric pressure, and were allowed to drop into the tail race, the head due to this elevation would be entirely lost, and in a plant operating under a low or medium head this would be a serious source of loss. To avoid this the turbine is made to discharge into an air-tight "suction" or "draught" tube which delivers the water below the surface of the tail race. When the turbine is started up the rush of water ejects the air from this tube, and when steady flow ensues the pressure at the point of discharge from the turbine runner is less than atmospheric by an amount equivalent to the elevation of this point above

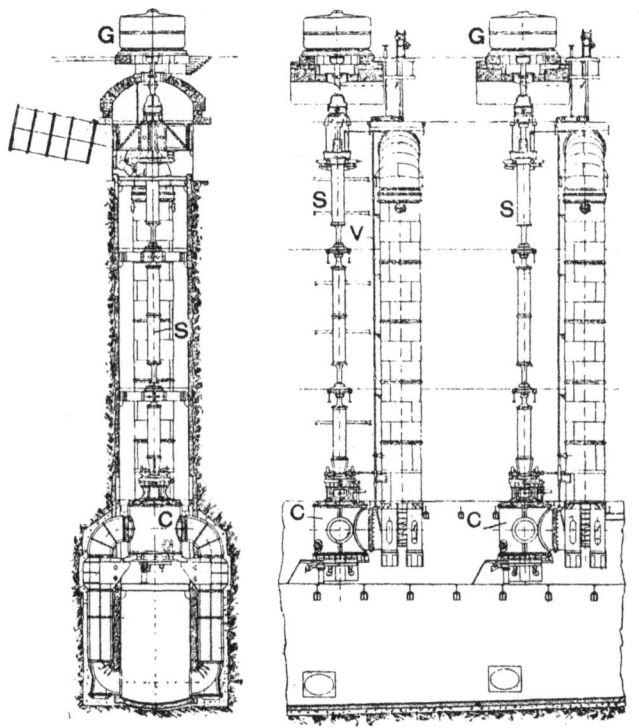

Fig. 16.   10,000 H. P. Turbines at Niagara Falls.

the free surface in the tail race.   The effective head

is thus the same as if the turbine, discharging freely into the atmosphere, were situated at tail race level.

The general arrangement of a turbine installation may be modified in many ways to suit circumstances. Where the head is low the turbine may well be erected in the open head race without any wheel casing.   Where the head is greater than about 40 feet, however, the necessity for making the wheel readily accessible renders a casing essential.   In some recent installations, notably that of the Canadian Niagara Power Co. at Niagara Falls, the turbine itself is placed at the bottom of a well of considerable depth and its power is transmitted, by means of a vertical shaft, to the electrical generating machinery which is placed above the level of the head waters. Fig. 16 shows the general arrangement of the turbines installed in the Power House of the above company. These machines have two runners on each shaft and develop 10,250 H.P. at 250 revs. per minute under a head of 133 feet.   Power developed in the casing C is transmitted through the tubular steel shaft S to the generator G.   Shaft S is 40 inches diameter and $\frac{9}{16}$ inch thick.   The total weight of the rotating parts is 120 tons and this is balanced by hydraulic pressure on the bottom face of the lower runner and on the lower face of a special balance piston keyed to the shaft.

Electric Generating Machinery.   Where the

electric current is to be utilized in the immediate
neighbourhood of the plant, or for electric railway
service, direct current generators may be used.
Under other circumstances alternating current, with
from 25 to 60 cycles per second, is more suitable.
The current is usually generated at about 2000 volts
pressure and is transformed up to the line voltage for
transmission.   The latter voltage may be anything
between 2000 and 100,000 volts, the higher the pressure the greater the power that can be delivered
over a given conductor.

Cost of Water Power Installations.   The cost of
a water power installation varies greatly with the
local circumstances and physical characteristics of
the site.   Where the available head is great, and
the storage reservoir is provided by some natural
lake, it may be comparatively small.   Where, on the
other hand, extensive works are required in order to
render the water available at the power house the
cost may be largely in excess of that of a steam plant
to give the same power.   An examination of some 120
European installations [20], most of which are supplying power for electrical distribution, shows that for
large installations of upwards of 10,000 H.P. the
minimum cost of the hydraulic works is £8. 8s.
per H.P. and the maximum £79. 11s. per H.P.   For
the majority of the installations this cost per H.P.
lies between £25 and £45.   The cost of the purely

electrical part of the installation also varies greatly, ranging from £1. 5s. to £28. 8s. per H.P., while the cost of the turbines ranges from about £4 to £8 per H.P. The working costs vary between £1. 6s. and £6. 16s. per H.P. per annum, with an average value of £3.

From these figures it appears that on the average, making an allowance of 15 per cent. for interest on capital and depreciation, the cost per H.P. per annum is in the neighbourhood of £10. 10s. With coal or oil fuel at its present price, this is largely in excess of the cost of energy developed in a large modern heat-engine plant. It is true that in many installations which are favourably situated the cost is much less than this. For example, at Niagara Falls power is generated at a cost in the neighbourhood of 15s. per H.P. per annum. Such installations are, however, the exception, and it may be taken as a general rule that at the present time heat-engine power is commercially more profitable than hydraulic power generated from the average installation, except in localities where the cost of fuel is abnormally high.

CHAPTER VIII

TIDAL POWER

IN virtue of its diurnal rotation about its own axis the earth possesses a store of kinetic energy whose amount is of the order of 3.7×10^{28} foot-pounds. It is impossible to make use of this energy directly, but since the effect of the rotation, combined with the gravitational attraction of the sun and moon, is to produce tides twice daily in the various bodies of water distributed over the earth's surface, any work which may be performed by tidal water during its rise or fall is due indirectly to the rotation, and must ultimately tend to reduce the velocity of rotation, and hence to increase the length of the sidereal day. Since, however, the length of the day would only be increased by about two seconds by the continuous abstraction of an amount of energy equivalent to 100 million H.P. for a million years, there would appear to be small reason for considering the effect of this on posterity.

In the simplest form of tidal power installation water is impounded in an artificial basin at high tide, and is allowed to escape into the sea through a series of turbines at and near low tide. Various modifications of this simple system are possible, and afford certain advantages.

If the whole of the water be stored until say one hour before low tide and be allowed to do the whole of its work during the next hour, the turbines will be working under a variable head having a mean value a little less than one-half the total tidal range. As the hour during which energy is being developed gradually works round the clock with the change in the time of low tide, it is evident that only on occasion will it occur during the ordinary working day, so that a storage plant, capable of absorbing the whole half-daily output, will be required as an essential part of such an installation.

In a second, and in some respects a preferable scheme, the turbines are allowed to work under an approximately constant head of about four feet. As soon as the level in the storage basin is say four feet above that of the ebbing tide, water is allowed to flow into the sea through the turbines, and the rate of flow is adjusted so as to keep the difference of level on the two sides of the turbine constant until low tide. Assuming a mean tidal range H of 20 feet, the level in the storage basin would thus only be allowed to fall through about 16 feet, and the available energy, with a given area of basin, would only be about one-third of that in the previous case. As against this the fact of working at constant head greatly simplifies the problem of speed regulation and tends to greater efficiency in the turbines, while, since the

installation is in operation for about five hours at
each tide, as against one hour in the former case, the
necessary capacity of any storage battery or accumu-
lator may be somewhat reduced.

Even with this system of operation, however, the
necessary capacity of such accumulators would be
considerable.   Power would be directly available
from the turbines for two periods of five hours each
daily, at intervals of about $7\frac{1}{2}$ hours, and assuming
a ten-hour working day, the slack interval would
from time to time be entirely included in the
working day.   The necessary storage system would
thus require to be capable of storing some 75 per
cent. of the daily demand.

With a duplicate system of turbines, arranged
to work under a constant head, one system being
operated by the inflowing tidal water, and the other
system as before by the outflowing water, the work
done per day with a given area of basin would be
practically doubled, and it would then be possible
to get useful work delivered from one or other series
of turbines for four intervals daily, each of about five
hours' duration and separated by idle intervals of
about $1\frac{1}{4}$ hours.   In this case the capacity of the
requisite storage plant would only require to be
sufficiently great to carry the plant over a $1\frac{1}{2}$-hour
period, or about 20 per cent. of that needed with
a single battery of turbines.

The necessity for large electric storage systems forms one serious drawback to either of the foregoing schemes, although this is not at all impracticable, even with the means presently available. A tidal power yielding 50,000 H.P. over a ten-hour working day, and having two five-hour working periods, would require a storage battery with a capacity of about 375,000 H.P. hours. This would probably be put up for about £550,000, on which, reckoning interest and depreciation at 15 per cent., the annual charge would be £82,500, or £1. 16s. per H.P. per annum. With duplicate turbines the storage capacity would require to be about 75,000 H.P. hours, costing about £110,000, and the annual charge would only amount to about 7s. 6d. per H.P. per annum. Against this, however, is to be placed the first cost and upkeep of the additional turbines, so that on the whole the ultimate cost of the two schemes would not be widely different.

With the whole of the water discharged during the hour around low tide, more elaborate turbines would be needed to enable fairly constant speed to be maintained over the whole range of heads, and the cost of the necessary storage batteries would be some 33 per cent. more than with the two five-hour working period system. On the other hand, the requisite area of the tidal basin per H.P. developed would only be about one-third as great as in the latter case.

The necessity for any storage system may be obviated by the provision of duplicate tidal basins separated by a wall in which the turbines are placed. If the lower basin be allowed to communicate with the sea during the lower third of the tidal range while the tide is falling, and the upper basin during the higher third of the tidal range while the tide is rising, the upper level never being allowed to fall below $\frac{2}{3}H$ and the lower level never allowed to rise above $\frac{1}{3}H$, the available head over the whole tidal range varies between about ·55H and ·80H with a mean value of about $\frac{2}{3}H$.

If the area of each basin be A square feet, a volume of water equal to $\dfrac{AH}{3}$ cubic feet may be passed through the turbines during the ten hours or so comprising the working day, and assuming a mean head of 20 feet and a turbine efficiency of 80 per cent., this would give 3100 H.P. for each square mile of area. This might be increased by possibly 25 per cent. by allowing the variation of level in the tidal basins to equal $\dfrac{H}{2}$, but the advantage would be largely counterbalanced by the necessity for more costly turbines and by the greater difficulties of successfully regulating the speed.

The elimination of storage batteries in this scheme is, however, only obtained at the expense of a greatly increased tidal-basin area; and for a given horse-power over a ten-hour working day the combined area of the two necessary basins would be about four times as great as in the first of the preceding schemes 33 per cent. greater than in the second, and about $2\frac{2}{3}$ times as great as in the third scheme (duplicate turbines). A special consideration of the circumstances of any proposed installation is thus necessary to determine which of these methods of operation offers the greater advantages.

In the course of operations carried out to improve the navigation of the Seine some twenty years ago, it was found possible to make the necessary training walls near Honfleur[21] enclose two basins with a total area of 2500 acres. These basins were divided by a bank in which turbines were erected. The cost of the special works necessary, not including the training walls, worked out at £72,000. The tidal range varies from 10 to 26 feet, and this gives an available horse-power of 3400 at neap tides and 8800 at spring tides. Taking the smaller value as being constantly available, the capital cost is £21. 4s. per horse-power developed at the turbine shaft, while on the mean power developed the cost becomes £11. 16s. per horse-power.

Although the cost of the training walls would greatly increase these values, it is evident that in special cases this method of utilizing the energy of the tides is quite feasible, even with coal at its present price.

The only tidal powers to be taken seriously into account are such as exist in localities where the tidal range is large and where, owing to natural advantages, comparatively short embankments or sea walls may be made to cut off large sheets of water. Such localities are to be found among the deep narrow fiords of Norway and among the Scottish lochs, but probably the most noteworthy is the Bay of Fundy, with its forty-foot tidal range.

The inner extremity of this bay forms an almost land-locked basin, having an area of more than 400 square miles. The headlands at its outlet are less than three miles apart, and it is safe to say that through this narrow gap energy, equivalent to more than 100,000,000 H.P. hours, runs to waste during the ebb and flow of each tide. To utilize this would require an engineering feat more tremendous than anything yet attempted by man, but in years to come the game may be worth the candle.

Various other methods of utilizing the rise and fall of the water during the ebb and flow of the tide have been suggested, ranging from the compression of air in huge cylinders during the rise of the tide to the use of floats which, during their rise and fall, operate gearing by which dynamos are to be driven. A little consideration of the necessary cost of any such devices, however, shows that this would be altogether disproportionate to any possible return, while the necessity for constructing these so as to withstand the enormous forces brought to bear on them during a storm appears generally to have escaped the notice of their inventors.

The energy of wave motion. Much ingenuity has been expended in the attempt to devise a successful wave-motor to develop useful energy by the rise and fall of the waves. In one device the rise and fall of a float carrying a plunger working in a stationary cylinder is made to compress air at each upward stroke, and various elaborations of this idea have been suggested. But of all the attempts to utilize natural sources of energy on any large scale, this must always be the most hopeless, not only from the fact that for weeks on end such an apparatus may be idle for lack of motive power, but also because of its enormously great cost per horse-power developed. However great may become the demand for energy in the future, it is difficult to see the slightest prospects of utilizing the energy of the waves with success.

CHAPTER IX

WIND POWER

UNDER favourable conditions the windmill is the most economical of all prime movers, since it is at once fairly cheap in itself and receives its energy perfectly free. The great drawback to wind power is its irregularity and uncertainty, and it is only in the region of the trade winds that it may be considered as anything like reliable. Furthermore, the largest windmills are of comparatively small horse-power, and in view of the massive and costly construction which would be necessary to enable a large mill to withstand such forces as might be brought to bear upon it during a gale, it is extremely unlikely that this prime mover will be built in much larger units than are in use at the present time. Calculations show that a unit to give an average output of 1000 horse-power would require a wheel of about 300 feet diameter, and such a plant would probably cost at least twenty times as much as one designed for 100 horse-power. Moreover, such a wheel, because of its comparatively large frictional resistances, would certainly be unable to take so great advantage of the lighter breezes which form so great a proportion of the whole, as would one of a lighter construction.

8—2

Fig. 17.   Annular Wind-Wheel driving a Scoop Wheel
for fen drainage.

The average velocity of the wind is low.   In most
places it is between 5 and 10 miles per hour, corresponding respectively to wind pressures of about
·125 and ·5 lb. per square foot.   Inland a 10-mile
breeze usually blows for about 50 per cent. of the
time, and a breeze of more than 5 miles per hour
for about 75 per cent. of the time.   In favourable
localities inland a 16-mile breeze may be expected
for about one-third of the year, while in some exposed positions in the neighbourhood of the sea
coast the average for the whole year is as great as
16 miles per hour.

Generally speaking velocities of less than 10 miles
per hour are of little use, while velocities of over
20 miles per hour necessitate sail area being reduced
in order to prevent excessive speeds of rotation.

Unfortunately, too, the windy periods do not
occur at regular intervals, and wherever the plant
be situated calms of several days' duration must be
provided for.   It follows that, if required to give
a constant supply of energy, any wind-power installation must include provision for storing a
comparatively huge proportion of its total output.

Proposals have been made to erect batteries of
windmills driving dynamos and to store the electrical
energy in secondary batteries until required.   A
rough estimate, however, shows that the cost of such
a storage plant to store a fortnight's supply would

be in the neighbourhood of £140 per horse-power, and that the annual charges and depreciation would amount to some £26 per horse-power.   When to this is added the first cost and annual charges of the windmill itself, it is evident that in spite of the free supply of wind energy the nett cost of utilizing this would be enormously in excess of the cost of steam, gas, or water power.

It has also been proposed to store the surplus energy by utilizing it to pump water into a reservoir or series of elevated tanks from which it may be allowed to flow through a turbine when required. With a convenient reservoir at hand this method need not be at all costly, but if it were necessary to construct a special reservoir or to build tanks the nett cost per horse-power would be little if any less than with electrical storage batteries.

Prof. Fessenden (British Association, 1910) has suggested sinking a shaft about 1000 feet deep, excavating a turbine chamber and reservoir at the bottom, and utilizing this, instead of an elevated reservoir, as a storage system.   For a plant to develop 1000 horse-power for say 60 hours per week would require a storage capacity of roughly 90,000 cubic yards for each week's supply.   Assuming a fortnight's storage and a down shaft 9 feet in dia-meter, this would involve some 190,000 cubic yards of excavation, which, even at the low estimate of

8s. per cubic yard, given by Prof. Fessenden, would
amount to £76,000, or £76 per horse-power. When to
this is added the cost of the turbine and its appur-
tenances, including a dynamo, it is evident that this
system of storage would necessitate a capital cost in
the neighbourhood of £100 per horse-power.

A further difficulty in the utilization of wind
power arises from the difficulty of speed regulation
in any exact degree. If the velocity of the wind is
doubled its pressure on any exposed surface is quad-
rupled, and hence with a windmill driving a constant
load the speed is considerably increased. With constant vane angles the speed of rotation for maximum
efficiency varies directly as the speed of the wind,
and under these conditions the output varies as the
cube of the speed. Thus, doubling the speed of
the wind increases the output eightfold. Generally,
owing to the comparatively high frictional resistances
of the mill itself, the power available for doing work
with a light breeze is comparatively small, and an
increase in the velocity of the wind increases the
output of the wheel in a greater ratio than even
the cube of the speed. Thus a well-constructed
windmill will just commence to rotate with a breeze
of 3 or 4 miles per hour, and will pump ten times
as much water with a ten-mile as with a five-mile
wind. On the other hand, the speed of the mill cannot be allowed to increase at the same rate as that

of the wind for very high velocities, so that at these velocities the working becomes inefficient. Thus the ordinary annular-disc wind-wheel will only pump some 50 per cent. more water with a breeze of 15 miles per hour than with a breeze of 10 miles per hour, while an increase up to 20 miles per hour only increases the discharge by a further 25 per cent.

To take the fullest advantage of the varying velocity and pressure of the wind the work done by the wheel should increase as the cube of the speed of rotation. In the reciprocating pump the power increases only slightly more rapidly than the speed, and considerably better results might be obtained for fairly large powers at high speeds by utilizing the wheel to raise water by means of a centrifugal pump. In this case the power increases slightly more rapidly than the square of the speed with a given pumping head. Alternatively the wheel might be coupled to a dynamo designed to give constant voltage over a fairly wide range of speed.

On the whole, although windmills have their uses as adjuncts in a general scheme of power supply, their utilization of the energy of the winds on any considerable scale has little to commend it. Like the earth's internal heat there is abundance of it as a whole, but little that can be utilized efficiently.

CHAPTER X

CONCLUSIONS

FROM a consideration of the facts and figures quoted in the foregoing pages, it is evident that to our present knowledge, even when the fossil fuels are exhausted, ample supplies of energy, renewable year by year, will remain for all the conceivable activities of the human race. Nor, indeed, in view of the gradual perfection of processes, is there any reason to anticipate that the average cost of such energy, taken the world over, will be very largely in excess of that of the fuel energy of to-day.

For the present, and for some considerable time to come, the fossil fuels will, in the majority of our present centres of activities, provide the cheapest and most convenient sources. These will be supplemented to an ever-increasing extent by the output of our rivers and waterfalls, whether near or remote from such centres. As, owing to the diminishing supplies of easily won coal- and oil-fuel, the cost increases, it would appear that the steam plant with its wasteful utilization of this energy will gradually become obsolete, except for very special purposes, and will be replaced by the more economical gas producer with its gas engine or turbine. Plants developing comparatively huge powers will probably

be installed in the immediate vicinity of the coal-
fields, and will generate energy to be transmitted
electrically to convenient centres of industry, while
the nitrogen of their fuel will be recovered and
returned to the ground in the form of salts of
ammonia.

As the cost increases still further, the considerable
peat deposits of the globe will be utilized, along with
the poorer bituminous shales of which such immense
deposits are found in the United States.

During the whole of this time, however, the stimu-
lus of increasing prices will have led to the extensive
development of power from vegetable sources, from
solar energy, and, where convenient, from the tides
and to a small extent also from the winds. Probably
it is safe to say that energy equivalent to that of
coal at something under £3 per ton could, under
favourable circumstances, be produced from any of
these latter sources. As each is successively tapped,
it will automatically check the demand for, and the
increase in the price of coal, and since, at our present
rate of consumption, the world's coal-fields would
probably last at least 2000 years, there is a strong
probability that even with any likely increase in the
general demand for power in the future, such fuel
will remain an important factor in the energy scheme
of the globe for at least this length of time. Ultimately of course its exhaustion is inevitable ; but

there can be nothing of the nature of a sudden change in such social or industrial conditions as would be affected by such exhaustion.

Little by little the hydraulic powers of the world will be pressed into service; acre by acre, fresh areas of the earth's surface will be brought under the intensive cultivation of such vegetable products as will yield the maximum output of fuel-value; solar plant after solar plant will be installed in regions adapted to their needs; some of these will probably transmit their energy to distant localities favourable to industrial life; others will enable what are now the waste places of the earth to be irrigated and to support human activity, and around them will gradually grow centres of industry: until when the last of the fossil fuel is consumed its loss will probably not be greatly felt.

But although mankind as a whole may not have suffered greatly by the change, and indeed may have benefited in some important respects, such an entire change cannot take place without having an enormous effect on the general distribution of activity on the earth's surface; and in the process isolated communities must inevitably suffer. The coal strikes of recent years afford a striking lesson of the suffering and dislocation of the social system which is produced in such a community as our own by even a temporary shortage of the coal supplies. And while

on the whole there is no reason for pessimism, or
for doubt as to the possibility of maintaining the
aggregate state of activity of the world, there is a
great deal of room for doubt as to the effect of these
changes on the state of our own and similarly situated
communities.

In this, as indeed in the majority of the industrial
communities of Europe the water powers are, in the
aggregate, of trifling amount; the climate renders
the direct use of solar energy impracticable; the
areas which might be utilized for the growth of
timber or of other vegetation for power purposes are
inadequate to provide fuel for more than an incon-
siderable fraction of the industrial requirements;
and in short the prospects of maintaining our present
supremacy as an industrial community in competition with other regions more favoured by nature, are
remote.

It is true that in our own case much might be
done by the importation of comparatively cheap fuel
—such as alcohol—while the prospects of supplying
Southern Europe with power transmitted electrically from solar installations in the arid regions of
Northern Africa are not unreasonably small.   But
in any case the handicap will be heavy.   Whether
sufficiently severe to counterbalance the advantages
of a temperate climate conducive to the development of the most energetic side of man's nature

remains to be seen. Burdened as we are by the necessity for a huge supply of fuel merely to enable existence to be maintained in comfort during a large portion of each year, the probability indeed is that it will be so, and that the coming ages will see a gradual drift of the industrial centre of gravity of the world towards warmer climes where, at all events, the necessity for artificial heating for domestic purposes will be largely non-existent.

Under the impetus of such a movement world-wide social and political changes will of necessity occur. Those regions favoured with an abundant supply of natural energy must gradually attain an importance altogether out of proportion to their present status, and an empire controlling such countries as Egypt, South Africa, Australasia, India, Burmah and Canada would appear to have every reason to rest content with the possibilities of the distant future. The United States also, with its South American neighbours, is not likely to suffer relatively by the change. The chief sufferers, from the point of view of national importance, are likely to be those northern industrial nations with no appreciable colonial possessions, and in view of this, the strenuous endeavours of such nations to obtain a due share of the sun-washed regions of the world become more than ever natural and intelligible. It would be true poetic justice if those regions from which originated the earliest civilizations of mankind came to be the centres of the latest civilization of all, and in spite of the vast changes which this would involve it is far from being impossible, or even improbable, that such will be the case.

REFERENCES

1. *Man and the Earth.* Schaler. N. Y. 1905.

2. *Liquid and Gaseous Fuels.* Vivian B. Lewes, p. 291.

3. *Phil. Mag.* April, 1907.

4. Presidential Address, British Association, 1911.

5. 1 B.TH.U. is the heat required to raise the temperature of one pound of water, at 60° Fahrenheit, through one degree F. The foot-pound is the work done in raising a weight of one pound through a height of one foot.

6. *Absolute Temperature.* If a gas such as air is heated or cooled under constant pressure, its volume is increased or diminished by a certain definite fraction ($\frac{1}{493}$) of its volume at freezing-point (32° F.) for each degree F. rise or fall. If this could continue down to $-461°$ F., *i.e.* 493 degrees below freezing-point, its volume would be reduced to zero, and this temperature is called the absolute zero of temperature on the Fahrenheit scale. Temperatures above this zero are called absolute temperatures and are got by adding 461° to the temperature as read on a Fahrenheit thermometer.

7. See Chapter II, p. 23.

8. H. A. Humphrey. *Proc. Inst. Civil Engineers*, Session 1912–13.

9. *Engineering News*, N. Y., Dec. 5, 1912, p. 1062. Also *Proc. Inst. C. E.* 1911–12, Pt. III, p. 427.

10. *Engineering News*, May 13, 1909, p. 512.

11. *Engineering News*, Sept. 21, 1911, p. 327.

12. By courtesy of Messrs The Sun Power Co., Ltd.

13. *Engineer*, Oct. 11, 1912, p. 393. As a result of experiment it has been decided to replace these zinc heaters by others of cast iron.

14. *Proc. Inst. Mechanical Engineers,* 1912, No. 1, p. 187.

15. *British Association Report,* 1889, p. 288.

16. *Engineering News,* Vol. lxv, 1911, p. 590.

17. *Trans. Inst. Mining and Metallurgy,* 1905-6, p. 455.

18. It is estimated that sufficient heat is conducted from the interior of the earth to its surface per annum to melt a continuous layer of ice, 4 ins. thick, over its whole surface. This is equivalent to a continuous outflow of energy amounting to about 1280 horsepower per square mile.

19. *Trans. Amer. Soc. C. E.* 1892, Vol. xxvii, p. 253.

20. *Zeitschrift für das Gesamte Turbinenwessen.* Munich and Berlin, 1908, pp. 357–362.

21. *Le Génie Civil,* Vol. xvii, 1890, p. 130. Also *Proc. Inst. C. E.* Vol. cii, p. 338.

BIBLIOGRAPHY

Fossil Fuels.

Coal. J. Tonge. Constable.
Geol. and Mineral Resources of New Brunswick, 1907.
Ibid. Summary Report for 1908.
Ibid. ,, ,, ,, 1909.
Liquid and Gaseous Fuels. V. B. Lewes. Constable.
Power Gas Producers. P. W. Robson. Arnold.
Report of Royal Commission on Coal Supplies, 1871.
Ibid. 1905.
General Reports of Mines and Quarries (Home Office).

Solar Heat.

Engineering News, N. Y., May 13, 1909, p. 512.
Ibid. Sept. 21, 1911, p. 327.

Interior Heat of Earth.

Cassier's Magazine, Feb. 1903, p. 566.
Brit. Assoc. Report, 1889, p. 35.

Water Power.

Hydraulics. A. H. Gibson. Constable.
Water Power Engineering. Mead. Wiley.
Modern Turbine Practice. Thurso. Constable.

Wind Power.

Proc. Inst. Civil Engineers, Vol. cxix, p. 321.
Ibid. Vol. cxlv, p. 387.

General.

Brit. Assoc. Report, Section G, 1910.
Ibid. President's Address, 1911.
Proc. Inst. C. E. Vol. cli, p. 18.
Report of British Science Guild, 1911.
Man and the Earth. Schaler. N. Y. 1905.

INDEX

For EU product safety concerns, contact us at Calle de José Abascal, 56–1°,
28003 Madrid, Spain or eugpsr@cambridge.org.